GENDER AND PHYSICS IN THE ACADEMY

Theory, Policy and Practice in European Perspective

Edited by
Pauline Leonard, Meytal Eran Jona,
Yosef Nir and Marika Taylor

BRISTOL
UNIVERSITY
PRESS

First published in Great Britain in 2024 by

Bristol University Press
University of Bristol
1–9 Old Park Hill
Bristol
BS2 8BB
UK
t: +44 (0)117 374 6645
e: bup-info@bristol.ac.uk

Details of international sales and distribution partners are available at bristoluniversitypress.co.uk

© Bristol University Press 2024

British Library Cataloguing in Publication Data
A catalogue record for this book is available from the British Library

ISBN 978-1-5292-2230-2 hardcover
ISBN 978-1-5292-2231-9 ePub
ISBN 978-1-5292-2232-6 ePdf

The right of Pauline Leonard, Meytal Eran Jona, Yosef Nir and Marika Taylor to be identified as
editors of this work has been asserted by them in accordance with the Copyright, Designs and
Patents Act 1988.

Cover design: Liam Roberts Design
Front cover image: iStock/virtualphoto
Bristol University Press uses environmentally responsible print partners.
Printed and bound in Great Britain by CPI Group (UK) Ltd, Croydon, CR0 4YY

FSC
www.fsc.org
MIX
Paper | Supporting
responsible forestry
FSC® C013604

Contents

List of Figures and Tables

Figures

Tables

1

Introduction

Pauline Leonard, Marika Taylor, Meytal Eran Jona, and Yosef Nir

This collection asks why the marginalization of women in physics remains endemic and systematic in European academic institutions. The issue is one of global concern: in a world struggling with health and climate crises, military and economic uncertainties, and widening levels of social and political inequalities, the critical roles that scientists play in both producing and mitigating safety and sustainability problems have never been more evident. Scientists are now publicly relied upon to identify risks and respond accurately to overcome global challenges, with physics playing a fundamental role in enhancing understanding about the world in which we live and thereby improving the quality of our everyday lives. Indeed, the influence of physics is far reaching, working together with other disciplines to develop technology, occupying positions with power to direct funding, and being influential in public debates concerning sustainability issues. Due to the status and power of physics in Western society, problems arise, therefore, if the scientists themselves are not drawn from the full range of diverse backgrounds. From a range of different theoretical foundations, academics and activists have demonstrated how women's marginalization impoverishes society. Some argue that diverse science is better science, and that including women's full contributions to discovery, knowledge production, and leadership are essential to achieve breadth, depth, and innovation in understanding. Others argue that distributing power more widely across scientific communities will help to challenge structures of knowledge and disrupt the imagination of future possibilities, both for physics and beyond. However, greater diversity not only enables inclusion of minority positions and a richer range of experiences and perspectives. The arguments consolidate to demand social justice, leading to more effective policymaking to benefit both science and humanity.

Undeniably, some significant gains have certainly been achieved towards gender equality within many academic disciplines. However, substantial challenges remain, not least within science, as well as technology, engineering, and mathematics (STEM). Further structures of inequality exist between and within STEM subjects, with physics particularly notable for low participation rates of women, and their poor representation in more senior roles. The Institute of Physics (IOP)'s member database shows that in the UK there is a large over-representation of those who identify as male (78 per cent) in comparison with those who identify as female (22 per cent) and compared with the gender breakdown within the general population (which sits at around 51 per cent females and 49 per cent males). Many European countries have similar or even poorer ratios (Eran Jona and Nir, 2019) with the percentages remaining consistent for over a decade. Further, despite the introduction of various policy initiatives over the last 20 years to improve the recruitment and career attainment levels of women in physics, the prospects of improvement are grim. Estimates suggest that if current hiring practices and attrition rates are maintained, the fraction of women in physics will remain below 30 per cent for at least 60 years. Even in the most optimistic alternative scenario, gender parity will not be achieved for another 25 years (Kewley, 2021; Leonard, 2021).

In this book, we draw on a distinguished team of leading physicists and sociologists from across Europe to take a comparative, interdisciplinary approach to investigating the extent of women's under-representation within the discipline and to enquire what can be done to improve their situation. Importantly, in a context where biological explanations are often posited to explain gender differences through factors such as lifestyle decisions or aptitudes (Stewart-Williams and Halsey, 2018), the book draws on social science perspectives to understand the reasons for low levels of women's participation, both in terms of entering the field and sustaining a career over the life course. We recognize the multiple differences which exist between women to understand gender intersectionally, as an aspect of identity which operates in a nexus including race and ethnicity, nationality/regionality, class and social background, religion, age, sexuality, caring responsibilities, and chronic illnesses and dis/abilities; all of which relate to power relations in different ways. To search for effective solutions for improvements in inclusion, we also explore various policy interventions adopted across different European contexts and evaluate their success. These theoretical and policy accounts produce the social architecture within which women physicists conduct their everyday working lives. To bring these experiences further to life, the collection is completed by inviting readers into the world of physics through personal life stories and visual images of women physicists in their own working environments.

The interdisciplinary nature of the chapters in this collection reflects the multidisciplinary nature of our editorial team. We are also ourselves a mix of physicists and sociologists from a range of social and international backgrounds. The idea for the book was first formed on a windy night in the Negev Desert in 2019, as we ate dinner and gazed at the stars, a few months before the world was shaken by the COVID-19 pandemic. We had come together through a gender and physics workshop organized by Yosef, Marika (both physicists), and Meytal, an organizational sociologist and gender expert, who had all been working with the gender and physics challenges for some years. Pauline, also a sociologist of gender and diversity in work and organizations, had been invited to give a talk at the workshop on sociological perspectives to the issue. Our joint passion for tackling inequalities due to differences in social identities was the fuel to the initial concept. On reaching out to many of the workshop participants for their views on the need for such a collection, it became clear that motivation for such a book was high, not only important as a resource but also providing an opportunity to evaluate experiences and achievements in this field.

We have spent much time debating the primary focus of the book: gender. While we fully acknowledge this is only one of the organizing principles of social life, it remains a fundamental axis of inequality in physics, not only in terms of numbers but also experience. The attention to gender is also important to us as editors of the collection and the book is underpinned by a feminist approach, presenting the debates through a diverse combination of research papers, theories, methodologies, and biographical forms. In this introductory chapter, we set the scene for the focus and the collection of chapters which follow. The chapter is divided into four sections. In the first section, we present a brief historical overview of gender and physics across Europe and beyond, identifying 'key moments' whereby women's presence and contribution have become recognized, and acknowledging some of the controversies that continue to the present day. Europe provides an important context for consideration of this issue, with many countries having recently introduced policy initiatives, both within individual nations and in consortia with others. The chapter then progresses to present an overview of social theoretical approaches to understanding gender inequality to situate the content of the book. Drawing on feminist sociological theory, the contributions of liberal, structural, and postmodern perspectives will be outlined. Building on this overview, section three reviews examples of different policy approaches to tackle gender discrimination, analysing how these are related to the different epistemological frameworks. The final section introduces the structure of the collection, linking this to the preceding discussion.

Gender and physics in Europe: historical background

While physics is traditionally considered a male discipline, there have been notable women physicists in Europe throughout history, dating back to ancient times. For example, Hypatia of Alexandria (370–415 CE) was a natural science philosopher who made significant contributions to astronomy and to mechanics. As head of the Neoplatonist school in Alexandria, she was considered one of the leading scholars of her time. In the Enlightenment period, important contributions to physics were made by women such as Laura Bassi (1711–78), the first woman physics professor in Europe, and Emilie du Chatelet (1706–49). While du Chatelet is perhaps best known for her translation of Newton's *Principia* into French, she developed new conceptual understanding of energy and momentum. In the 19th century, a small number of women published influential books on physics such as *The Mechanism of the Heavens*, written by the astronomer Mary Somerville (1780–1872). Women scientists in these periods were typically privileged socio-economically, with sufficient family resources to pursue their intellectual interests.

During the first half of the 20th century, women remained under-represented in physics in Europe. While this is clearly evident from the absence of women in historical accounts of physicists who became significant in the discipline, the invisibilization of women is exacerbated by the complete lack of acknowledgement of their role as assistants, wives, partners, and so on. The support work provided by women to male protagonists is slowly being written back in to historical accounts, in part to rectify the Herculean notions of physics which positions men as brilliant 'stars' who operate independently of those around them. This point is well argued in later chapters of the book. Our discussion here of women who gained recognition for their research in physics is thus underpinned by our appreciation of all the work that women perform 'under the radar'.

In 1912, Ernest Solvay founded the International Solvay Institute for Physics, which organizes the celebrated Solvay Conferences on Physics. In the early conferences, physicists from all over Europe gathered to discuss the revolutionary new theories of physics, quantum theory, and relativity, but the only woman participant in these meetings was Marie Curie (1867–1934), who later became famous for her work on radioactivity and twice a winner of the Nobel Prize.

Women scientists faced challenges in obtaining academic posts and recognition for their work. Lise Meitner (1878–1968) was part of the team that discovered nuclear fission, but her contributions were often overlooked due to her gender and nationality (Austrian). Cecilia Payne-Gaposchkin (1900–79) made the groundbreaking discovery that stars are primarily made up of hydrogen and helium but faced barriers throughout her career. She completed her undergraduate studies at the University of Cambridge but did not graduate because the university did not grant degrees to women until 1948. She relocated to Harvard University for graduate studies as her only career path in England was teaching. At Harvard she spent many years in low-paid, low-prestige research posts because the university did not grant professorships to women; she only became a full professor in 1956.

During the 1960s and 1970s, European universities experienced significant expansions. In the same period, the labour-market participation rate for women increased, but the numbers working in physics remained low. For example, it was not until 1971 that the UK appointed its first woman full professor in physics, the nuclear physicist Daphne Jackson. Daphne Jackson campaigned for women's rights and in 1985 she devised a plan to support women to restart their careers after a career break. The Daphne Jackson Trust was founded after her death in 1992 to provide fellowships to scientists who have taken career breaks of at least two years for caring or health reasons.

In the 1990s the continued under-representation of women in physics was explored in several influential reports. In 1992 the IOP published a report titled 'Women in Physics: A Survey of Employment and Training in the UK'. This report was based on a survey of over 700 women physicists, as well as interviews with employers and professional bodies. It found that women were significantly under-represented in senior academic and industrial positions. For example, data from the UK Higher Education Statistics Agency demonstrates that even by the late 1990s only 1 per cent of physics professors were women. The 1992 IOP report identified a range of factors contributing to this under-representation, including lack of support, bias, and stereotyping, and proposed recommendations such as promoting positive role models, addressing bias, and increasing support and funding for women. Since the 1990s the IOP has actively engaged in addressing these issues, working in collaboration with UK universities and industry, as well as international partners. The role of the IOP in accrediting university physics departments for their work on gender equality is discussed in Chapter 9.

A second influential publication was the 1999 Massachusetts Institute of Technology (MIT) report 'A Study on the Status of Women Faculty in Science at MIT' led by Nancy Hopkins (Hopkins, 2002). This study was based on surveys and interviews with staff and students in science departments at MIT. Many women staff members reported feeling marginalized and excluded from significant roles in their departments. Data revealed that this marginalization was often accompanied by differences in salary, space, awards, and resources. The percentage of permanent staff (8 per cent) was found to have not changed significantly in the preceding decade. The MIT report made a number of recommendations for addressing these issues and the Dean of Science took immediate actions to effect changes resulting in a rapid increase in the number of women in the staff. The MIT report drew attention to the challenges faced by women in science and had considerable impact both nationally and internationally.

Following these reports, a number of European countries launched 'women in physics' initiatives. In the UK, the Royal Society launched the Dorothy Hodgkin Fellowship scheme in 1995, aimed at supporting postdoctoral scientists needing flexibility due to caring responsibilities or personal health reasons. Although men can apply for the scheme, the type of flexibility offered particularly meets the needs of women; the scheme offers maternity leave, part-time working, and assistance with childcare costs during conferences and visits to collaborators. As described by Wilkin and Miller-Friedmann in Chapter 9, the UK also launched its Athena SWAN and IOP Juno accreditation schemes in this period. These encouraged higher education institutes to apply for 'badges' to recognize good practice and achievements in improving gender equity. Over the passage of time, as the importance of gaining such awards has become more significant for university branding, so the levels of self-assessment and equality, diversity, and inclusion initiatives have increased.

In the Netherlands, the physics research council FOM launched a dedicated programme to support women in physics in 1999 (De Hoogh et al, 2019). This programme had flexible funding that could be used to support the retention of women in physics, either through postdoctoral positions or through bridging funding to enable physics institutes to offer permanent posts to women. Typically, around two women per year were awarded funding and many of the alumni of this programme have progressed very rapidly to senior positions. Several universities in the Netherlands have permanent track fellowship schemes for women, such as the Rosalind Franklin scheme in Groningen, launched in 2002. While such positions are not specifically for women in physics, these schemes have attracted many talented international women physicists to the Netherlands.

In 1999 the European Commission published a document 'Women and Science: mobilizing women to enrich European research' (Commission of the European Communities, 1999). This report led to the formation of the European Platform for Women Scientists (EPWS) in 2003, which received funding from the EU under the Sixth and Seventh Framework Programmes for Research and Technological Development. The EPWS now represents over 12,000 women researchers with more than a hundred member networks in 40 countries. Later in this book, we will explore other European programmes, such as the Gender Equality Network in the European Research Area (GENERA) project, that specifically focus on gender in physics. These have enabled continuity of the focus on women's participation in the discipline, although the commitment to improving gender imbalances remains partial and inconclusive across different national contexts.

In 2003 the European Commission published its first report of data on women in science: 'She Figures: Women and Science: Statistics and Indicators'. These data are now published every three years, and follow the path of women from graduating doctoral studies to participating in the labour market and acquiring decision-making roles while exploring differences in women's and men's working conditions, and research and innovation output (European Commission, 2013). She Figures reports enable tracking of trends, facilitating identification of fields of science and technology in which progress remains slow. For example, the percentage of women in the highest academic positions in physics is increasing but only at a rate of around 1–2 per cent over three years.

Turning beyond Europe now to the global situation of women in physics, the International Union of Pure and Applied Physics (IUPAP) founded a working group on Women in Physics (WiP) in 1999. The IUPAP is a global organization which currently has 60 member countries; it works closely with national and regional physics societies in America, Africa, Asia, and Europe. The WiP working group in the IUPAP monitors gender data and promotes actions to increase inclusion and diversity. One of the key actions of the working group is the organization of a triennial global conference on gender diversity in physics, the first of which was held in 2002 (Hartline, 2002). The proceedings of these conferences are an important source of data about the situation of women in physics around the world.

The IUPAP WiP group led global surveys of women physicists in 2002 and 2006, with male physicists included in a third survey carried out in 2009–10. In 2017–19 the IUPAP worked together with 11 other international science unions to investigate the gender gap in STEM globally and across disciplines, and published a detailed report (Gledhill et al, 2019). The report included findings from a global survey with 32,000 participants, investigation of gender impacts on publication patterns, and a compilation of best practices across the world.

The outcome of these various reports, research studies, working groups, and policy initiatives has undoubtedly been to raise awareness of gender issues, and the multiple challenges facing women building careers in academic physics departments in Europe. However, while the consistency of activity over the last 30 years to keep gender on the agenda has meant that discussions for improvements in both participation and experience continue to be ongoing, at the same time this also indicates that success has been partial and fragmented, and that the issues facing women persist within the discipline. That this book argues for change after all the work and time that has been put into the issue is somewhat depressing. It is clear that there is still much to be done to achieve a more balanced picture, which demands understanding and tackling the reasons behind the deeply embedded structures of inequality. The contributions of sociology are helpful here, as we now turn to consider.

Gender inequality and physics: sociological and feminist perspectives

Sociological theoretical perspectives on gender and inequality offer a valuable resource to understand physics as a site of work, conducted in departments and organizations where gender difference, disadvantage, and discrimination pervade. While national differences clearly exist within European academia, the chapters in this book also reveal strong structural and cultural similarities within higher education institutions. A diverse body of groundbreaking research and theoretical development over the last 20 years has established a range of frameworks to understand the complex relations between gender, work, and organizations. For overviews, see, for example, Halford and Leonard (2001), Konrad et al (2006), Jeanes et al (2011), Lewis et al (2019), and Tyler (2021). As a heuristic device, we can cluster these approaches into three broad perspectives: liberal, structural, and postmodern, while recognizing that the boundaries between these are often blurred to produce more nuanced perspectives and arguments.

The most popular/dominant way of understanding the relationship between gender and organizations draws on 'liberal' traditions, which rest on assumptions of the underlying sameness of women and men's abilities, to argue for gender equality and justice (Halford and Leonard, 2001; Finlayson, 2016). In this perspective, women and men are assumed capable of the same behaviours and achievements, not least in the capacity for rationality (to think and act with reason and without being swayed by emotion). Differences in skills and feelings are mapped onto women and men socially, such that 'gender' is constructed through self-sustaining distortions such as sex role socialization and stereotyping, prejudice, and discrimination, all of which work to privilege men and marginalize women through the allocation of resources, status, authority, and power. Liberal perspectives offer an account for the pervasiveness of gender discrimination across society; a series of patterns caused by a range of individual and social preconceptions and prejudices about gender, often unconsciously. Organizations such as academic institutions form key arenas where gendered beliefs, biases, customs, and prejudices continue to operate: the male professor who derides women as unable to take the pressures of hard science and prone to emotional outbursts; the PhD supervisor who asks women students about their plans for motherhood; the line manager who assumes that giving birth equates to declines in ambition. As we see from the contributions in this book, such attitudes continue to thrive in physics departments in European academia.

Structural perspectives rest on the belief that social relations between individuals, in organizations as well as elsewhere, are part of a broader, deliberately orchestrated, structured system of power relations between unequal social groups based on gender, class and/or race, and so on (Halford and Leonard, 2001). Rather than reflecting the isolated actions of prejudiced individuals, these structures systematically underpin social and economic life, created in the interests of dominant groups (for example, White/middle class/men) to perpetuate their privilege and power. Work organizations are a key vehicle for reinforcing these patterns of structured domination: many, if not most, are controlled by male senior management teams, with the gender wage gap and temporary contracts being key economic manifestations of the structural oppression of women. University departments, such as physics, represent an 'ideal type' of structured gender inequality, with male physicists routinely concerned to preserve their domination of the discipline.

The structural perspective is further subdivided by variations in the broader argument. Radical feminist approaches, in contrast to liberal assumptions of similarity between women and men, emphasize the *differences* between women and men in mind and body; for example, focusing on 'men's and women's different ways of leading' (van der Boon, 2003). Material feminists claim that, historically, men organize as a social group to dominate women: a systemic oppression termed patriarchy. They enquire how this set of power relations has come about and urge that attention is paid to how historical social relations in different parts of the world shape the present (Basham, 2021). They demonstrate how 'the political, economic, juridical and ideological have worked with and through each other to substantiate patriarchal social relations that often normalise the spatial separation of social spheres' (Basham, 2021: 178). Within this collection, Lund and Aarseth argue for a revisiting of material historical feminism, valuable as a lens to better understand academic physics as a world of separate spheres. In recent years, a different form of 'feminist new materialism' has emerged, aiming to break down ontological barriers between 'environments', 'objects', and 'bodies' (Barad, 2007; Gough and Whitehouse, 2018). New material feminisms also shift the emphasis away from the human as the centre of social relations to propose 'that all manner of bodies, objects and things have agency within a confederation of meaning making' (Taylor, 2019).

Socialist/Marxist feminism also considers that women as a social group are oppressed, but positions capitalism as the primary cause for this structured inequality (Armstrong, 2020). Women form an essential 'reserve army' of labour to be drawn upon or rejected according to the needs of profit, with their primary function being the reproduction of the labour force. This perspective might be traced in the report 'Women and Science: mobilizing women to enrich European research', which was referred to previously (Commission of the European Communities, 1999). The boundaries between radical and Marxist feminism are also being blurred within new material feminism, to produce broader definitions of materiality which acknowledge embodied aspects of race, sexuality, imperialism and colonialism, and anthropocentrism (Alaimo and Hekman, 2008) as structuring mechanisms.

Black feminism critiques both radical and material feminist approaches to argue that neither adequately recognize race and racism. The concepts of 'gender' and 'class' have been constructed as essentially White categories which do not adequately account for the experiences of Black women (Love, 2016). Racism is a structural feature of society in which Whites as a social group systematically dominate and subordinate Blacks as a social group. Bureaucratic institutions such as universities are essentially White forms of organization, established to serve the interests of White people.

These interconnections between forms of oppression due to, inter alia, race, class, and gender are captured in the concept of 'intersectionality' that explores the ways in which different forms of social inequality, oppression, and discrimination interact and overlap in multidimensional ways (Crenshaw, 1989; Lutz et al, 2011). The concept demands feminist thinking not only integrates the full range of marginalized perspectives but also demonstrates the necessity of understanding relations of rule and power differentials as co-constituted and co-constitutive. This necessitates looking at the different social positioning of women and men, and the diverse ways in which they participate in the reproduction of these relations (Lutz et al, 2011). In this investigation, postmodernism and poststructuralism have been particularly productive. Shifting away from understanding gender as a largely dualistic, constitutive attribute (Tyler, 2021), a more performative ontology is proposed, which emphasizes the ways in which gender is the outcome and not simply the basis of organizing processes. Gender and race are performances we *do*, not attributes we *have* (West and Zimmerman, 1987; Rutherford, 2020): 'routine accomplishment[s] embedded in everyday interaction[s]' (West and Zimmenrman, 1987: 125). As a social practice, gender is both multiple in its performances and provisional in its meanings (Fancher, 2016; Twine, 2018), differentially constituted in specific contexts of time and space. One of the principal contributions of postmodernism, therefore, is the critical focus on any assumption of stability in meanings about gender, such as are implied by the terms 'women' and 'men'. In other words, performances understood as traditionally 'masculine' or 'feminine' can be undertaken by anyone, regardless of their gender. However, it is also acknowledged that while a wide variety of alternative constructions about masculinity and femininity exist, such that predictions of behaviour are non-foundational, performances of gender are not free-floating. Rather they are compelled by powerful discursive regimes, such as 'the heterosexual matrix' (Butler, 2010) and the specificity of specific historical and social processes.

Consciously or unconsciously, these epistemological frameworks inform the regimes of meaning within which explanations of gender inequalities are produced; in turn, informing different visions of policy interventions. However, it is rare that this is a 'neat and tidy' relationship where one perspective clearly feeds into a specific policy. Rather, policies often reflect a complex basket of approaches and assumptions, as we now turn to explore.

From theory to policy

Different theoretical perspectives lead to different understandings of the methods by which to tackle and solve gender inequalities successfully. For example, working from a liberal perspective, it is commonly argued that unconscious bias, prejudiced attitudes, and outmoded stereotypes prevent women from achieving their full potential at work. This is not only 'bad for women' but also 'bad for organisations', which need to benefit from the full talent pool of human capacity (Carosella, 2020). Within this perspective, the organization itself is seen as a gender-neutral entity, which becomes patterned by the prejudices of certain individuals within it to impede gender equality. What is needed, therefore, are clear rules, processes, and procedures that will remove or counter people's biases and outcomes based on gender prejudice. Effective policies are the means by which visions of institutional neutrality and meritocracy can be achieved. At the core of this approach is the belief that women and men are capable of the same levels of competency and should therefore be enabled to be equal in terms of pay, status, ambition, experience, and so on. An equal and just organization is one which deploys effective policies to ensure this.

Structural approaches take a more fundamental approach to policy. For radical/material feminists, a key plank is for organizations to acknowledge the benefits of 'women's ways of knowing': approaches to practice which draw on women's superior abilities with empathy, sensitivity, communication, and compassion (Belenky et al, 1986). In contrast to styles which in Western society are often associated with masculinity, such as those which draw on rationality, objectivity, independence, and calculation (the epistemology of conventional science) (Alimo-Metcalfe, 1994), women's knowledge contributes to distinctly different and, for some, even richer, forms of understanding. For socialist and Black feminists, full equality can only be achieved through pay and other material benefits. Removing pay gaps based on inequalities in gender, class, and race are the only effective means by which differences in material outcomes are to be addressed. Other policies include full equalities in parental leave, state-provided care, and pension rights (Day et al, 2021).

While postmodern approaches eschew policies through the understanding that no 'one-size-fits-all' approach can tackle the multiple complexities involved in gender/power relations, at the same time it is recognized that challenges to the gendered basis of power relations are important. 'Performing equality' and disrupting the meanings of 'femininity' and 'masculinity' through our performances and interactions are means by which power can be distributed differently, albeit that this awareness is needed in each and every social interaction (Mavin and Grandy, 2012; Greig, 2021). Policies which are sensitive to the needs and inclusion of particular groups and sections of society, such as Black women PhD programmes, can be regarded as in the spirit of postmodernism. New materialism joins the disturbing of binaristic boundaries by adding in recognition of, inter alia, the body and the sensory, agreeing that it is through a full and diverse range of disruptive practices in performance that power relations can be challenged (Coleman et al, 2019). Policies on acknowledging the menopause by work organizations have been influenced by such perspectives.

Against this epistemological backdrop to contemporary thought and policymaking on gender inequality, we now turn to introduce the chapters in this book. We identify the positions taken by our contributors, and the similarities and differences in terms of how they understand gender, gender inequalities in physics, and effective policies to work for change.

Structure of the book

The chapters in this collection are authored by academic researchers and policy makers from a broad range of disciplinary and national contexts, providing a rich and diverse compendium of theoretical understanding, topical empirical research, and experiences of policy interventions. While this multiplicity of interpretations and policy recommendations means that finding an all-purpose solution may feel near impossible, the depth and range of the contributions provides a valuable armoury of knowledge and strategies to improve understanding of the complexity of issues involved in tackling gendered inequalities. We start, in Part I, with two chapters which offer theoretical accounts to better understand gender disparities in physics. These are extended, in Part II, by three chapters which draw on theoretical frameworks to analyse the findings of new empirical research which variously investigates women's experiences as physicists. The two chapters in Part III each explore major European policy initiatives: the GENERA project, an EU initiative which ran between 2015 and 2018; and the UK's IOP's Juno project. In Part IV we turn to explore the lives and experiences of women physicists in personal detail through in-depth interviews and photographs submitted by a range of international scholars to depict their working contexts. Finally, in Part V we draw the contributions together in our concluding chapter, where we discuss the recommendations for best practice which emerge from the collection.

Part I

In Chapter 2, Tomas Brage and Eileen Drew argue that both structural and cultural factors need to be understood before policies can be designed to change the deeply gendered nature of both higher education institutions and the physics departments within them. Drawing on the legacies of radical feminism, they describe how physics has perpetuated a 'Herculean' masculinist culture which is, at best, chilly and, at worst, hostile to women. This is based on (false) notions of a 'culture without a culture' (Traweek, 1992) in physics: the assumed consequence of a science based on positivistic/ masculinist assumptions of objectivity. Digging even deeper into the roots of this powerful discourse, Brage and Drew uncover the heritage of the Christian Church and ancient trades to describe how powerful masculinist metaphors of 'priests of knowledge' and 'blacksmiths' still linger in imaginations of 'the ideal physicist'. Women have long been positioned as oppositional to such cultural identities and, at the same time, are subject to structural patriarchal oppression through recruitment malpractices and institutional discrimination. The authors argue that solutions must lie in both structural and cultural institution-wide transformation to challenge the epistemologies that dominate not only physics but also much of the academy.

In Chapter 3, Rebecca Lund and Helene Aarseth argue that the neoliberal restructuring of higher education institutions in the past 20 years has led to a 'remasculinization' of the academy. This operates in subtle and complex ways, such as the emphasis on standardized metrics of 'excellence' which ignore the local and contingent factors affecting individual performances. In an innovative retrieval and reconceptualization of feminist historical materialism, the authors demonstrate how energies within the academy are split along gendered lines to fundamentally differentiate women and men's careers. As inherently capitalist institutions, the division between abstract and concrete labour persists to underpin the gendered sub-text of contemporary neoliberal universities. As a social group marked by lower levels of pay and contractual security, women continue to be those who invest more time and energy in caring relations, such as teaching, mentoring colleagues, and participating in equality and diversity interventions, while men, positioned as most are in higher status and secure roles, are more likely to construct their energies through more conventionally recognised standards of academic 'excellence'.

Part II

In Chapter 4, Meytal Eran Jona and Yosef Nir draw on a theoretical combination of structural and liberal approaches to explore a key stage of the career trajectory: the transition from PhD to postdoctoral researcher. This is often a 'critical moment' of decision-making as to whether to remain in the academy or pursue a science career elsewhere. The chapter offers theoretical novelty in conceptualizing career decision-making as a 'deal' which involves intersections between contextual, organizational, and individual variables which are framed within an overarching gendered power structure. Within this power structure, gender operates in multiple and often hidden ways to construct barriers to women's academic careers and thereby influence women's experiences and decision-making. Drawing on empirical mixed-methods research conducted in Israel, the authors reveal how the academic field is male-dominated and unequally competitive, while the social and domestic spheres position women pursuing an academic career as disruptive to the gender order of Israel. However, their research reveals that a love of physics and a deep interest in the field are also powerful motivators, as well as senses of freedom and enjoyment of the working conditions that research delivers. They conclude that the 'deal' for women has three components: personal-marital; professional-occupational, and financial. The decision whether to leave or remain is made by individuals through a consideration of these multiple factors – a limited form of agency which is bounded by a deeply gendered structural context.

Themes of structure/agency, as well as the performative aspects of 'doing gender', are developed further in Chapter 5. Drawing on qualitative interviews conducted with women physicists across 12 European countries, Paulina Sekuła explores experiences of discrimination through 'gender micro aggressions': subtle insults which take place during everyday interactions that communicate a hostile and/or dismissive attitude towards women and, in the process, produce an invalidating climate for them. Microaggressions are a means by which male power can be executed as a consistent drip-feed, and the detrimental effect on women's psychological well-being, self-confidence, comfort at work, and productivity only serves to further impede their professional development and career progress, already impacted by discrimination through recruitment and promotion practices. The result is that women must find strategies to cope with microaggressions – a process which is both exhausting and frustrating. That at present the solution remains with individual women to manage their own situations reveals the lack of accountability by institutional hierarchies to remedy their deeply embedded power structures.

These themes are further investigated in Chapter 6, where Meytal Eran Jona and Yosef Nir turn to discuss a second research study on the experiences of women researching for their PhDs in Israel. Taking a broadly liberal feminist approach, they discuss how the academic context of physics in Israel is marked by deeply gendered assumptions of women's roles. These not only present women with work–life conflicts, but also routine incidents of microaggression and sexual harassment. Their investigation drew on a large-scale survey of all the physics PhD students in Israel to gain a deeper understanding of the under-representation of women in order to formulate policy recommendations to improve gender balance in Israeli physics. Their study revealed three main areas of challenge for women: professional, economic, and personal. While these challenges are faced by all PhD students regardless of gender, women also experience gender-specific hurdles which are hidden to the academic system: 'glass hurdles' of psychological and physiological health problems; pregnancy, motherhood, and family care and the discrimination which results from this; as well as ongoing sexual harassment. The chapter concludes with some policy recommendations which are strongly liberal in form: to remove these 'glass hurdles' and thus improve the situation for women physicists.

Part III

In the three chapters of this section of the book, we turn to take a comparative approach to considering contemporary policy approaches to gender inequalities in physics. Across Europe, countries both together and separately have been introducing a range of interventions to try and address the long-standing structural and cultural issues patterning the discipline. In Chapter 7, Thomas Berghöfer, Helene Schiffbänker, and Lisa Kamlade review the GENERA project, an EU initiative which ran between 2015 and 2018, designed to develop inclusiveness and establish physics as a more gender-balanced field. The project aimed to take what the authors term a 'bottom-up' approach, involving women and men physicists on all career levels in the design of Gender Equality Plans (GEPs) to improve inequality within the discipline. This involved data collection to identify gender bias and strategies to address this, such as family-friendly policies, training on gender equality, and setting targets to monitor progress. Underpinning these priorities are liberal assumptions that improving individual behaviours and ameliorating barriers to progression presented by factors such as childcare will provide a more egalitarian context in which men and women can compete on equal terms. No attempt was made to dismantle the deeper structural inequalities marking the discipline, although targets were identified to evaluate change. That, by the end of the project, not all GEPs had been signed off by the respective institutions, is indicative of the (often subtle) resistance initiatives such as these can meet. While the authors conclude that change takes time, and that progress is never linear, interesting questions are raised about the effectiveness of liberal approaches to effect significant structural transformation.

In Chapter 8, Yolanda Lozano and Marika Taylor describe the work carried out by an EU-funded project – European Cooperation in Science and Technology (COST) – to address the particularly large gender imbalance in string theory. Lozano and Taylor commence by drawing on their own experiences as string theorists to review the culture of the field, and the gender dynamics within this. The chapter exposes some deep structural patterns to women's participation in string theory, where covert assumptions about the level and form of women's inclusion resonates with Marxist understandings of marginalized and precarious labour markets. The authors then turn to explain the origins of the COST project and its objectives to improve awareness and understanding of the gender imbalance, and to promote explicit actions to address gender equality. The project collected quantitative and qualitative data which highlighted gender issues in the environment and culture of the string theory research community. The chapter concludes by discussing the ongoing impacts of the project on the string theory research community.

Chapter 9 forms a bridge between Part III and Part IV, by exploring the UK's Athena Scientific Women's Academic Network (SWAN) and the IOP's Juno project both in relation to the schemes' achievements and personal experience. In a fascinating and innovative format, the chapter interweaves a contextual and political overview of the two schemes (written by Jaimie Miller-Friedmann, a science educationalist) with Nicola Wilkin's auto-ethnographical reflections on her own career as a physicist in the light of these. In these honest accounts, suggestive of radical feminist approaches to the inclusion of women's voices, the chapter reveals that while the schemes have undoubtedly delivered personal benefits to the UK's women scientists, and Nicola's own career, these are not without some substantial costs. The fact that it is mainly women that become involved in implementing equality policies can also detract from their career building within physics. The chapter raises important issues about the 'burden of representation' which is an outcome of the low participation rates of women in the discipline, yet which results in heavy demands for speaking 'as a woman', with all the complexities this poses. The observation that constructing gender as a fundamental difference within policies designed to improve women's experiences and outcomes can mean that being a 'woman' physicist becomes a primary identity is especially pertinent, serving both to *empower* women while, simultaneously, maintaining their *disempowerment*.

Part IV

In Part IV, we turn to represent women physicists more personally and individually. In Chapter 10, Marika Taylor presents a series of biographical interviews held with four physicists: three women and a man who has been committed to improving gender equality within the discipline. Their stories reveal the diversity in life courses and attitudes towards, and experiences of, gender inequality. This theme is taken up further in Chapter 11, where we give women voices and faces. In this chapter we offered women physicists to present photographs of themselves in their chosen work environment, with a short description of why and where the image was taken and what is shown. The images, edited by Meytal Eran Jona, are deeply personal but also convey powerful messages of pride and comfort in their surroundings.

Chapter 12 closes our collection with our concluding thoughts. While there have been some considerable efforts to improve gender equality within physics, change has been slow, and improvements are significantly smaller than in other scientific disciplines. Drawing on the evidence of best practice revealed by the latest research in this area, concerted, strategic, and bold leadership is now needed for effective structural change, such that evidence-based policies and successful role models can drive forward greater equity. This book aims to provide conceptual tools and empirical evidence for this urgent task.

References

Alaimo, S. and Hekman, S.J. (eds) (2008) *Material Feminisms*, Bloomington, IN: Indiana University Press.

Alimo-Metcalfe, B. (1994) 'Waiting for fish to grow feet! Removing organizational barriers to women's entry into leadership positions', in M. Tanton (ed) *Women in Management: A Developing Presence*, London: Routledge, pp 27–45.

Armstrong, E. (2020) *Marxist and Socialist Feminism*, Study of Women and Gender: Faculty Publications, Northampton, MA: Smith College.

Barad, K. (2007) *Meeting the Universe Halfway: Quantum Physics and the Entanglement of Matter and Meaning*, Durham, NC: Duke University Press.

Basham, V. (2021) 'A necessarily historical materialist moment? Feminist reflections on the need for grounded critique in an age of crises', *International Relations*, 35(1): 178–82.

Belenky, M.F., Clinchy, B.M., Goldberger, N.R., and Tarule, J.M. (1986) *Women's Ways of Knowing: The Development of Self, Voice, and Mind*, vol 15, New York: Basic Books.

Butler, J. (2010) *Gender Trouble: Feminism and the Subversion of Identity* (10th edn), New York; Oxford: Routledge.

Carosella, C. (2020) 'Why Gender Equality Matters in Business Success', *Forbes*, [online] 27 March, Available from: www.forbes.com/sites/forbesn onprofitcouncil/2020/03/27/why-gender-equality-matters-in-business-success/?sh=3244d24f669c

Coleman, R., Page, T., and Palmer, H. (2019) 'Feminist new materialist practice: the mattering of methods', *MAI Feminism and Visual Culture*, Spring Issue (3), Available from: https://maifeminism.com/feminist-new-materialisms-the-mattering-of-methods-editors-note/

Commission of the European Communities (1999) 'Women in Science: mobilizing women to enrich European research', Available from: http://aei.pitt.edu/13321/1/13321.pdf

Crenshaw, K. (1989) 'Demarginalizing the intersections of race and class. A Black feminist critique of anti-discrimination, feminist theory, and antiracist politics', *University of Chicago Legal Forum*, pp 141–50.

Day, M., White, C., and Kaur, S. (2021) 'The pay and progression of women of colour London', *Fawcett Society and Runnymede Trust*, [online] September, Available from: www.fawcettsociety.org.uk/Handlers/Downl oad.ashx?IDMF=c1300375-f221-4a88-8c66-edf3c30bd2c7

De Hoogh, A., Hesping, S., Rudolf, P., and De Wolf, E. (2019) 'The Dutch FOm/f approach to gender balance in physics', *AIP Conference Proceedings*, 2109(1): 050028.

Eran Jona, M. and Nir, Y. (2019) 'Women in physics in Israel: an overview', *AIP Conference Proceedings*, 2109(1): 050022.

European Commission (2013), *She Figures 2012: Gender in Research and Innovation: Statistics and Indicators*, Publications Office of the European Union.

Fancher, P. (2016) 'Composing artificial intelligence: performing Whiteness and masculinity', *Present Tense: A Journal of Rhetoric in Society*, 6(1): 1–9.

Finlayson, L. (2016) *An Introduction to Feminism*, Cambridge: Cambridge University Press.

Gledhill, I., Roy, M.F., Chiu, M.H., Ivie, R., Ponce-Dawson, S., and Mihaljević, H. (2019) 'A global approach to the gender gap in mathematical, computing and natural sciences: how to measure it, how to reduce it?', *South African Journal of Science*, 115(3–4): 1–3.

Gough, A. and Whitehouse, H. (2018) 'New vintages and new bottles: the "Nature" of environmental education from new material feminist and ecofeminist viewpoints', *Journal of Environmental Education*, 49(4): 336–49.

Greig, A. (2021) 'Doing gender differently; transforming masculinity', *UNDP*, [online] April, Available from: www.undp.org/blog/doing-gen der-differently-transforming-masculinity

Halford, S. and Leonard, P. (2001) *Gender, Power and Organisations*, Basingstoke: Palgrave.

Hartline, B.K. and Li, D. (eds) (2002) 'Conference resolutions', *AIP Conference Proceedings*, 628(1): 3–8.

Hopkins, N. (2002) 'A study on the status of women faculty in science at MIT', *AIP Conference Proceedings*, 628(1): 103–6.

Jeanes, E., Knights, D., and Yancey Martin, P. (eds) (2011) *Handbook of Gender, Work and Organisation*, Chichester: John Wiley.

Kewley, L.J. (2021) 'Closing the gender gap in the Australian astronomy workforce', *Nature Astronomy*, 5(6): 615–20.

Konrad, A., Prasad, P., and Pringle, J. (eds) 2006 *Handbook of Workplace Diversity London*, London: Sage.

Leonard, P. (2021) 'Women getting in and getting on', *Nature Astronomy*, 5(6): 533–4.

Lewis, P., Benschop, Y., and Simpson, R. (2019) *Postfeminism and Organisation*, London: Routledge.

Love, K. (2016) 'Black feminism: an integrated review of the literature', *ABNF Journal*, Winter: 11–15.

Lutz, H., Viva, M.T.H., and Lupik, L. (2011) *Framing Intersectionality: Debates on a Multi-Faceted Concept in Gender Studies*, Farnham: Ashgate.

Mavin, S. and Grandy, G. (2012) 'Doing gender well and differently in management', *Gender in Management*, 27(4): 218–31.

Rutherford, A. (2020) 'Doing science, doing gender: using history in the present', *Journal of Theoretical and Philosophical Psychology*, 40(1): 21–31.

Stewart-Williams, S. and Halsey, L.G. (2018) 'Men, women, and science: why the differences and what should be done', Available from: https://scholar.google.co.uk/scholar?hl=en&as_sdt=0%2C5&q=Stewart-Williams+and+Halsey%2C+2018+&btnG=

Taylor, C. (2019) 'Diffracting the curriculum: putting "new" material feminist theory to work to reconfigure knowledge-making practices in undergraduate higher education', *Turning Feminist Theory into Practice: Enacting Material Change in Education*, Dordrecht, Netherlands: Sense Publishers.

Traweek, S. (1992) *Beamtimes and Lifetimes: The World of High Energy Physics*, Cambridge, MA: Harvard University Press.

Twine, F.W. (2018) 'Technology's invisible women: Black geek girls in Silicon Valley and the failure of diversity initiatives', *International Journal of Critical Diversity Studies*, 1(1): 58–79.

Tyler, M. (2021) *Judith Butler and Organisation Theory*, New York: Routledge.

Van der Boon, M. (2003) 'Women in international management: an international perspective on women's ways of leadership', *Women in Management Review*, 18(3): 132–46.

West, C. and Zimmerman, D. (1987) 'Doing gender', *Gender and Society*, 1(2): 125–51.

PART I

Theoretical Perspectives

Embedding Gender Equality into the Culture and Discipline of Physics: Unfinished Business

Tomas Brage and Eileen Drew

Introduction

Research-performing institutions are important for promoting gender equality, diversity, and inclusion but, despite many positive changes, they remain deeply gendered institutions. The lack of gender parity at senior professorial and decision-making grades has adverse consequences for innovation across all disciplines. Physics has proved to be no exception; indeed, it represents a predominantly and consistently male-dominated discipline. Referring to the Nobel imbalance of 2021 awards, Sanderson (2021) noted that since the Nobel prizes were first awarded in 1901, there have been just four women[1] physics laureates, compared with 12 women physiology or medicine laureates and seven women chemistry laureates. Of similar concern is the lack of progress towards gender parity among those who aspire to study physics: 'girls make up only 20 per cent of A level physics students in the UK and 20 per cent of physics undergraduates in the US' (Cooper, 2020: 224). Statistics for Ireland indicate that, in 2017, women comprised 23 per cent of total degree, diploma, and certificate holders in physics across all higher education institutions (HEA, 2020). This chapter explores how the discipline of physics has evolved to ensure a perpetuation of a masculinist Herculean 'culture without a culture' (Traweek, 1992) requiring a concerted response from not only physics and physicists, but also from science, technology, engineering, and mathematics (STEM) faculties and higher education institutions themselves to drive gender sensitivity in academe.

The European Commission (EC)'s role has been central in seeking fundamental structural and cultural change in European academia, through projects such as Gender Equality Network in the European Research Area (GENERA), Institutional Transformation for Effecting Gender Equality in Research (INTEGER), Systemic Action for Gender Equality (SAGE), and GenderEX. The evolution of these EU initiatives and policies contextualizes and reinforces the need for *institutional* gender-equality policies and Gender Equality Plans (GEPs). As Heilman and Okimoto (2007) have stressed, it is women who face negative consequences, in traditional male-dominated domains (such as physics) whether they act in a gender-typical or atypical way, in attempting to conform to, or reject, masculine norms of behaviour. The role of both structural and cultural institution-wide transformation is increasingly recognized as necessary to underpin the gender dimension in physics and other maths-intensive fields, to include knowledge production and curricula, thereby contributing to gender sensitivity.

The discipline of physics is exemplified by many metaphors and superlatives: a culture of no culture, 'Herculean', and 'priesthood or blacksmith?', operating in a culture of 'effortless excellence' populated by a 'nerd culture of cosmopolitans' (Wertheim, 1997; Hasse and Trentemöller, 2008; Dippel, 2021). In order to counter this male-dominated culture, physics needs to be supported by gender-sensitive institutional processes and strategies, requiring a vision of gender equality that crosses disciplinary boundaries and engages with a variety of gender and feminist perspectives. Physics does not exist in a vacuum. The distancing from a binary idea of natural sciences versus humanities, as described in the agential realism of Barad (2007), provides a pathway to understanding the interplay of nature and culture in the knowledge production of physics. This chapter sets out a blueprint for action that includes institutional structural and cultural interventions that challenge prevailing behaviours, attitudes, cultures, and even the popular (among physicists at least) epistemology. Specific measures are required to create a more gender-sensitive discipline of physics thereby attracting a more inclusive, diverse, and gender-balanced quorum of students, researchers, academic staff, and decision-makers.

Background

Achieving gender equality is an increasingly important policy requirement for academic institutions. Faced with enduring gender inequalities among academics and administrators, at all levels, university leaders have been charged with defining action strategies to ensure the effective implementation of structural measures to reduce and eliminate gender bias in their organizations. Universities play a crucial role in promoting gender equality and diversity; the last 20 years have produced a range of positive changes, through an enlarged pool of highly qualified women in academia and the wider labour market. Notwithstanding these positive developments, progress has been uneven – not least in terms of discipline.

For more than two decades, the EC has charted the lack of retention of women at each level of the academic ladder, in which women comprised less than 10 per cent of the leaders in the 'scientific system', despite the fact that half the STEM graduates were women (European Commission, 2000). The European Technology Assessment Network (ETAN) report pinpointed the forms of discrimination, often unconscious, against women. It also identified the key problems faced by women in scientific careers: the flawed operation of the peer review system; and low levels of engagement by women in shaping scientific policy and setting the agenda in the top committees of the EU and of member states. The report advocated a sustainable improvement of women's standing in science and research, requiring a significant transformation of science and scientific institutions (European Commission, 2000).

Subsequent EC reports highlighted additional problems hindering the progress/achievement of gender equality. In particular, the Women in Research Decision-Making (WIRDEM) expert group report (WIRDEM, 2008) identified nomination procedures, cultural barriers, and funding limitations as hindering factors in the progress of women in their academic careers. It reviewed member states' policies and existing procedures for evaluating and promoting researchers to senior positions, outlining examples of good practice at national and institutional levels.

The current focus for EU-wide activities is on the research institutions and organizations where women in science work, *rather than just on the women themselves*, through 'fixing the administration' as the major objective (European Commission, 2010: 12). In response to these endemic gender-inequality issues, a European Research Area (ERA) survey pointed to actions that research organizations could take, such as recruitment and promotion measures, targets to ensure gender balance in recruitment committees, flexible career trajectories (for example, schemes after career breaks), work–life balance measures, and support for leadership development (European Commission, 2015). In this book you can read about a number of initiatives, supported by the EU, to enhance gender equality in physics (the GENERA project, Chapter 7, as well as other projects, see Part III).

It is increasingly acknowledged that progress is further impeded by the striking gender inequalities that persist in career advancement and participation in academic decision-making. Despite significant progress in their level of education over recent decades, relative to men, women are progressively under-represented as they move up the stages of an academic career, particularly in STEM disciplines, as depicted in the She Figures 2021 publication (European Commission, 2021).

The Understanding Puzzles in the Gendered European Map (UPGEM) report (Hasse and Trentemöller, 2008) noted with concern that well-qualified women scientists often leave the research system prematurely and those who stay rarely, or never, reach the top-level positions (grade-A professors) or achieve distinguished careers in research and development in the same way as their male counterparts. In the EC's report 'Structural Change in Research Institutions' (European Commission, 2012) it is argued that gender-aware management of universities and research organizations would have a positive impact on policies and practices in recruitment, promotion, and retention of both women and men, ultimately benefiting the quality of the research itself.

In 2012, the League of European Research Universities (LERU) issued recommendations for governments, funders of research, academic publishers, and, most notably, universities to address gender deficits through embarking upon actions that sought: commitment at the top and throughout the institution to gender equality; and development or implementation of a gender strategy and/or action plan with the support of all divisions and levels within the university.

More recently, it has been recognized that a third level of actions is needed, to analyse and engender knowledge production, thereby feeding into the curriculum for teaching and learning. The Gendered Innovation project sought a thorough inclusion of a gender and sex analyses for an inclusive science to ensure quality which may save resources and lives (Nielsen et al, 2017). LERU (2015) published an advice paper on how to integrate gender and sex in research and innovation. These resources are mainly concerned with research in STEM where sex (as in biological organisms) or gender (as a determinant of 'culturally acceptable' human behaviour) is 'obvious' in the subject of study. This could be in research that involves animals and cells, that are obviously sexed, or applied science – for example, in transportation, virtual reality, or car seat belts – that relate to a sex *and* gender dimension. This chapter addresses the more 'non-obvious' parts of STEM, where resistance to such a perspective could be significant – not least due to a lack of understanding and good practice examples in physics.

The next section of this chapter sets out the case for intervention within the discipline of physics, as a case study of a male-dominated science in which the prevailing culture is one that represents a chilly climate for women who cannot adapt to the 'Herculean' prototype. Drawing on a wide body of literature, it sets out the all too prevalent hostile academic environment that has evolved and been reinforced in the evolution of physics as a discipline. In conclusion, we address the issue of 'engendering physics' in its knowledge production and teaching, supported by a more gender-sensitive institutional environment.

What does gender have to do with physics?

This question implicitly raises a 'positivistic paradox' in which physics is grounded in an objective, genderless description of reality, based on empirical evidence from observations that are independent of the 'bodies and minds' of the researchers. Yet the history of the discipline, classroom norms, and particularly decision-making in physics, have been dominated by men. How is a subject that seems inherently independent of sex and gender so gendered in its culture? Schiebinger (2001) offers three dimensions to tackle this question in terms of fixing the numbers, culture, and knowledge. These dimensions are clearly intertwined, not least in physics, in which the prevailing culture defines what knowledge is worth searching for, even in 'curiosity-driven' basic science and *whose* curiosity drives the research?

Cold, hard numbers

When looking at Schiebinger's fixing the numbers, it is clear that physics, more than most other subjects, exemplifies both horizontal and vertical segregation (Schiebinger, 2001). Horizontal segregation means that women and men gravitate to different fields when choosing their careers. Usually, attempts to counter horizontal segregation consist of trying to convince women outside academia that 'physics is fun', or that they should take more maths courses in high school or similar measures. Without belittling these efforts, the 'changing the [numbers of] women' mentality is inherently flawed, by relying on 'fixing the women', unless actions are taken on the other Schiebinger dimensions: culture and knowledge.

Vertical segregation provides the key: the fact that men are promoted in academia at the expense of women, particularly in physics, which resides in the system itself. An often-used metaphor for such segregation is the 'leaky pipeline' implying that there is only one pipe to funnel through towards a successful career. In reality, things are more complex, and different people need different pipelines to thrive and stay in academia (Ong et al, 2017; Harvard Project). Second, the pipeline analogy suggests that people who leave academia have 'failed', yet these people often end up in successful careers in other fields (Etzkowitz and Rang, 2011). The drain of talent is therefore a problem for academia, not just for those who leave. Something within the academic culture of physics seems to repel women.

Culture of no culture

Schiebinger's (2001) dimension of culture has been studied by many disciplines including physics. In a classic anthropological study of major American and Japanese science labs, Traweek (1992) demonstrated how the social structure within physics is formed, how excellence is defined, and how young scientists are groomed. She argued that physicists consider (and want) their labs to be a 'culture of no culture', in which physicists are so convinced of the objectivity of positivistic science that they believe it also defines the way they interact personally and professionally. This is increasingly recognized as problematic, since there is evidence that a strong belief in objectivity and a working meritocratic principle, could lead to an even stronger subjectivity in selection processes (Castilla and Bernard, 2010). Traweek's groundbreaking work is a gold mine for anyone who would like to understand the culture of physics; for example, how new generations are groomed and the human relationship to machines and particles. Despite describing a culture from the 1990s, her work remains valid, since academic cultures change slowly, and it has inspired a wealth of later work.

Priest or blacksmith?

One interesting feature of physics is its historical close ties to religion – contrary to the preferred narrative of physicists (Wertheim, 1997). In their history of science, through textbooks and lectures, physicists refer to the noble battle of rationality against the powerful Church, thereby ignoring the co-dependence of physics and religion in the early years of the Enlightenment; for example, in how Newton and Kepler invoked the 'God of the gap' to explain inexplicable parts of their models, such as how the force of gravitation works on planets at a distance. The link between the Pythagorean idea of a world described, even ruled, by mathematics was construed as striving to understand God, a principle that is referred to even by the 'fathers of modern physics'. Using the metaphor of Pythagoras' trousers, Wertheim (1997) traced how physics grew from religion (while other sciences emerged from more 'practical' origins – for example, alchemy and early medicine). An example of an early perceived hero of physics, Giordano Bruno, is illustrative of the 'adjusted narrative' of the history of physics. Bruno was burnt at the stake by the Church not only for supporting the Copernican heliocentric world view, but also, more problematically for the Church, he believed in magic and mysticism – that the human mind could alter the faith of the world. Contemporary physicists were as opposed to these ideas as the Church (Wertheim, 1997). The idea of a physicist being a 'priest of knowledge', with special abilities, belonging to a select few, helps to explain the prevailing culture, which is often hostile to ethnic minorities and women. Hence there is a need to address this by introducing a gender perspective.

Another equally gendered metaphor is the physicist as a blacksmith, portrayed as an engineer who builds and runs machines (Vainio, 2013). Physics is an essential component of engineering in which building, 'mastering', and relating to machines are important. Even modern engineering, in the form of electronics, follows the same gender lines and can be alien to, and hence exclude, students that have not been grounded in physics from an early age. Faulkner's (2000) gender-aware research on engineering underlined the still-prevailing connection between masculinity and technology. The same also holds for programming that has been hijacked by men. It is clear that both the priesthood and blacksmith metaphors reflect highly masculine concepts.

Debunking Hercules

The UPGEM project (Hasse and Trentemöller, 2008) was a sociological study undertaken in five European countries (involving Italy, Denmark, Poland, Finland, and Estonia). It investigated and assessed the reasons underlying the stark country-to-country disparities in the percentages of physics professors who were women, ranging from 23 per cent in Italy to 3 per cent in Denmark, linked to the culture of the countries – for example, childcare, career choices in science at secondary school level, religion, and status of the professoriate. However, these did not provide the full explanation. The researchers did find a correlation between the lack of women at the highest academic levels and how strongly 'Herculean' the institutional cultures were. The Herculean idea of a single, strong leader who successfully advances science defines this culture in Denmark, while in Italy there was more of a 'caretaker' attitude in which team building and networking was considered important for the production of excellent results and good leadership. It is implied that the Herculean culture cannot benefit anyone who does not fit that reductive stereotype.

Hence it is time to redefine and abandon the model of a brilliant, single scientist, since it fails in modern physics where problem-solving groups and collaboration are the norm. Moreover, diversity in building such groups is necessary and the key to excellence (Nielsen et al, 2018). Dippel (2021) argues that physics has started to change by moving towards a 'nerd culture of cosmopolitans', combining an almost introverted love for science, with a broad and international network of scientists.

The meritocracy myth

At the heart of the issue of retaining women in physics is implicit bias (LERU, 2018; Gvozdanović and Bailey, 2021). According to Wennerås and Wold (1997: 343), 'studies have shown that both women and men rate the quality of men's work higher than that of women when they are aware of the sex of the person to be evaluated, but not when the same person's gender is unknown'.

Such bias represents a threat to the principle of meritocracy – even as research shows that the idea of meritocratic universities itself might be a myth. In a recent study of Aarhus University in Denmark, it was found that 20–30 per cent of professorships were appointed in closed processes and around 40 per cent of the rest had only one applicant (Nielsen, 2015; Nielsen, 2016). At the same time, the percentage of women appointed through open competition was double that in closed processes. It is clear that bias is not only an unconscious or implicit cognitive process, since the cultural context affects our decisions. Systemic bias is embedded through the decoupling of actions from the stated meritocratic and academic principles (Nielsen, 2021). Bias can arise in the utilization of apparently objective 'metrics', where bibliometric and quantitative standards can mask underlying preferences (for example, numbers rather than quality of outputs). A third example of systemic bias is symbolic boundary work, where stereotypical arguments are used to explain the slower careers for women, while presenting somewhat cloudy ideas of differences between men and women regarding, for example, their risk-taking and caring capacities versus competitive natures.

Biases, compounded by the meritocratic myth, impede the achievement of an open and transparent career path in physics. Castilla and Bernard's (2010) research shows that the more convinced a group is that it follows meritocratic principles, the more likely it is affected by bias. Since physics is defined by a strong meritocratic idea in its devotion to objectivity it might be in danger of being more affected by bias in evaluations, promotions, and recruitment processes.

Particularly threatening to gender equality and diversity is the issue of harassment and discrimination. A study from Uppsala University, Sweden (Lundborg and Schönning, 2007), interviewed doctoral physics students and found that around one third of the women had experienced sexual harassment at work and/or at conferences. At Chalmers University of Technology in Gothenburg, Sweden (Berg et al, 2012), 53 per cent of women interviewed had experienced harassment on the grounds of their sex or gender. A more recent study of doctoral physics students found that women experience extra challenges, compared with men, most notably the 'glass hurdles', including sexual harassment, gender-based discrimination, and more (see Chapter 6, Eran Jona and Nir). An important and more subtle part of the problem comes in the form of 'micro-violence' – those everyday actions in the form of being ignored, made invisible, ridiculed, or withheld information (see also Sekula's study on women physicists' experiences of microaggressions in Chapter 5).

Beyond the 'objectivity' defence

Drawing upon Schiebinger's (2001) dimension of knowledge, particularly of 'the subject', instead of holding fast to the idea that physics is purely objective, she suggests that there are aspects of science that can be improved by embracing a gender perspective. A definition of physics that only includes logical discussions, equations, and formal mathematics is no longer tenable. While electrons and equations have no sex/gender, in real life, physics bleeds into other fields and deserves a broader definition. Physics is what physicists do, and that involves a wide variety of gendered practices.

Physics research does not happen in a vacuum – it must be contextualized in terms of how it is performed, its purpose, and who benefits from it. Similarly, a physics teacher needs to use examples and metaphors and select topics that are interesting and important. Clearly, physics is affected by the background of the researcher, teacher, and student, and it follows that a gender perspective is needed. It can also be framed in the words by Wertheim: 'we must also re-examine the longstanding perception of physics as a "transcendent" pursuit for, as I have noted, this view continues to act as a major cultural barrier to women' (Wertheim, 1997: 247). And earlier she notes: 'Western culture has evolved conceptions of "science" and "femininity" as polar opposites. … This polarization is especially acute in the "hard" science of physics. … Since the prevailing conception of femininity is the opposite of the "blunt, brilliant bastard," there is a problem here' (Wertheim, 1997: 247).

As in other fields, physics has also been discussed from a feminist epistemological and even ontological perspective. Barad (2007) argued for 'agential realism', based on the Copenhagen interpretation of quantum mechanics. This offers an alternative to the extreme positivism of Newtonian mechanics, without decay into relativism, in which the reality, experiments, and observer are entangled, thereby opening up a clear and deep gender perspective of physics with the focus on addressing imbalance, cultural change, and engendering of research and the physics curriculum.

The prevalence of physics as embedded in masculinity reinforces the need to apply a gender perspective to challenge the apparent adherence to objectivity in the discipline and the perpetuation of a culture that is exclusionary.

Effortless success or hard work?

The recent anthology edited by Gonsalves and Danielsson (2020) provides several examples of how identity research can be used as a lens to understand how gender is intertwined with physics education. A study by Archer et al (2020) followed women on their route to the academy to detect what could repel them or exclude them from continuing to study physics. In school, boys and girls tend to be equally attracted to the method of physics – the empirical science with a mathematical description of the phenomena – but they are interested in different applications. This implies that we can make physics more inclusive, not by changing its core ideas but in how we apply them.

It was also evident that the identities of the young women did not match with the typical identity of physics and they had to renegotiate them – a task that requires effort and energy, which might affect the performance of women. A special problem was the often-mentioned idea of what makes a successful physicist – a person who does physics with 'effortless success'. In this context, both teachers and students gave the impression of success coming naturally and easily to some, and these people will be the true physicists, thereby echoing the idea of priesthood. In reality, success comes from hard work, not just talent. For Jammula and Mensah (2020), physics is sometimes paradoxically portrayed as being without entry points: you cannot learn physics if you don't understand it! The door closes for anyone who does not recognize themselves as one of the select few.

Arising from these elitist arguments, Stewart and Valian (2018) emphasized that the feeling of not belonging, particularly for minoritized groups, leads to anxiety. In turn, this can cause reduced performance and a strong feeling of not fitting in, thereby creating a vicious circle. To counteract this, it is necessary to debunk the myth of effortless success. Physics is more or less effortless depending on the 'social capital' you bring with you in the form of (for example) academic parents, quality of high school education, and expectations. Work by Archer et al (2020) shows that a simple thing, such as the choice of examples, can make or break the feeling of belonging – physics is too often represented by examples and metaphors from the world of sports, cars, and even super-heroes. Acknowledging that it is common to feel anxiety and that it can change seems to make a difference and can improve performance and retention (Stewart and Valian, 2018).

Resistance to gender mainstreaming in physics education often comes in different forms. First, a questioning of its importance (the 'objectivity myth') – this can be countered by referring to the wealth of studies on physics and gender. Second, it could be argued that there is no room for a gender dimension in the curriculum or that the teachers do not have the right preparation for it. This shortcoming could be overcome by education and awareness-raising on the importance of learning social aspects and outcomes of physics. A working strategy would be to introduce academic teacher training and modules for students. The latter will often create a demand for further actions and the final goal of the integration of a gender perspective into regular courses.

Why physics 'solo run' initiatives cannot succeed without an institution-wide GEP

Uncovering and acknowledging that anomalies exist is an important stage in seeking to redress gender imbalances and improve the academic culture, in the context of achieving excellence in teaching and research. This section draws upon some of the noted practices that impede women's entry, retention, and progression up the academic career ladder, notably in recruitment and workload allocation.

Recruitment malpractices

Research shows that some department heads mobilize cultural narratives about the different characteristics and attributes of women and men to account for academic successes and failures in their departments. Consequently, symbolic boundary work contributes to legitimizing persistent gender inequalities in hiring and selection. Nielsen (2021) concluded that:

> Denmark is a compelling example of the importance of ensuring transparency and continuous oversight in all recruitment and selection activities. Gender bias thrives in situations where recruiters can exploit loopholes to make recruitment and selection run 'fast, cheap and smooth'. Universities need to close these loopholes with the help of administrators and gender equality taskforces. Universities should ensure, through monitoring, that appointments are made based on fair and open procedures and that recruiters are held accountable when this is not the case. (Nielsen, 2021: 37)

This importance attributed to recruitment practices underlies the need to ensure open and transparent procedures *at an institutional level*, and monitoring to indicate or highlight any anomalies that need to be investigated and rectified. Once appointed, women may still face adverse conditions relating to uneven workload allocations, bullying and harassment, and unconscious bias. Hence, even if faculties and physics departments/schools seek to rectify gender imbalances in appointments, they may encounter resistance and adherence to a masculine status quo in the institutional recruitment/selection process.

Lessons from EU projects

Important lessons have come out of EU-funded institutional change projects that involved the engagement of physics (such as GENERA and INTEGER, see Chapter 7), not least in the strong emphasis on developing and implementing institutional GEPs. First, it is crucial to avoid gender binaries and to address intersectional diversity. Second, actions must be: data driven, designed to be gender-neutral, tailored to the micro (for example, departments of physics) or macro environment (the higher education institution itself) in which they operate; and strategically designed to align with the institution's core values. Third, it is important to involve men as champions/active participants in the transformational change process. Fourth, actions need the support from top level as well as key allies – for example, human resources and equality, diversity and inclusion (EDI) offices – to influence gender-related policies and overcome resistance. Fifth, it is important to apply an active communication strategy, conveying information on actions and their benefits to the widest possible community of stakeholders. Sixth, while responding to the local context and specific needs, acting at institutional and school/departmental levels, making use of/extending existing training/development opportunities, GEPs should prioritize unconscious bias awareness training.

The SAGE project reiterated that gender-equality interventions and GEP formulation/implementation should be pursued at not just discipline level but also across the entire institution *and* supported at sector level by national authorities/funders (O'Connor and Irvine, 2020; Drew, 2022). This integrated approach requires changing the system to ensure congruity across the institution. In summary, the following actions would be required to enable the cultural transformation of physics:

- tackle institutional cultures by introducing bias-awareness training, bias observers, anti-discrimination workshops, and supporting teamwork over a 'Herculean' culture;
- create gender-integrated leadership and career programmes for all;
- investigate why research and academic staff leave, and take actions to counteract to facilitate retention, particularly when exit patterns are gendered;
- evaluate the effects of gender-equality awards and certification (such as Juno, Athena SWAN, and Gender Certification) with a view to adoption of an adaptable scheme;
- counteract horizontal segregation in STEM while avoiding approaches that only aim to 'change the women';
- embrace the adoption of GEPs at institutional and discipline levels, particularly for physics;
- introduce measures to engender research based on research and good practice; and
- integrate a gender equality and diversity perspective in the physics curriculum, starting with the training of academic teachers, followed by the introduction of special modules with the long-term goal of integrating this perspective into mainstream physics courses.

As part of the academic world, physics reflects the extreme end of positivism upholding, we would argue, an almost cult-like obsession with objectivity. Being the 'jewel in the crown' among the sciences, its Herculean culture is extreme, representing a closed society of excellent 'priests of physics'. Solving the 'positivistic paradox' for physics will involve meeting and countering resistance against change towards a more gender-inclusive academy.

Note
[1] Anne L'Huillier became the fifth woman to win the Nobel Prize in Physics.

References
Archer, L., Moote, J., Macleod, E., Francis, B., and DeWitt, J. (2020) *Aspires 2: Young People's Science and Career Aspirations, Age 10–19*, London: UCL Institute of Education.

Barad, K. (2007) *Meeting the Universe Halfway: Quantum Physics and the Entanglement of Matter and Meaning*, Durham, NC: Duke University Press.

Berg, C., Edén, S., Heimann, S., Peixoto, A., Silander, C., Turner, D., and Wyndham, A.K. (2012) *Jämställda fakulteter?*, Sweden: University of Gothenburg.

Buitendijk, S., Curry, S., and Maes, K. (2019) 'Equality, diversity and inclusion at universities: the power of a systemic approach', LERU PG EDI position paper, Available from: www.leru.org/publications/equality-diversity-and-inclusion-at-universities

Castilla, E.J. and Bernard, S. (2010) 'The paradox of meritocracy in organizations', *Administrative Science Quarterly*, 55(4): 543–76.

Cooper, Y. (2020) *She Speaks*, London: Atlantic Books.

Dippel, A. (2021) 'Ontological opportunism: reanimating the inanimate in physics and science communication at CERN', *Anthropological Journal of European Cultures*, 30(1): 27–51.

Drew, E. (2022) 'Navigating unChartered waters: anchoring Athena SWAN into Irish HEIs', *Journal of Gender Studies*, 31(1): 23–35.

Etzkowitz, H. and Ranga, M. (2011) 'Gender dynamics in science and technology: from the "leaky pipe-line" to the "Vanish Box"', *Brussels Economic Review*, 54(2–3): 131–47.

European Commission (2000) *Science Policies in the European Union – Promoting Excellence through Mainstreaming Gender Equality*, report from the ETAN (European Technology Assessment Network), Expert Working Group on Women and Science, Luxembourg: European Commission.

European Commission (2010) *Stocktaking 10 years of 'Women in Science' Policy by the European Commission 1999–2009*, M. Marchetti and T. Raudma (eds), ERA, Science in Society, Luxembourg: European Commission.

European Commission (2012) *Structural Change in Research Institutions: Enhancing Excellence, Gender Equality and Efficiency in Research and Innovation*, Luxembourg: European Commission.

European Commission (2015) *Strategic Engagement for Gender Equality 2016–2019*, Brussels: European Commission, Available from: http://ec.europa.eu/justice/genderequality/document/files/strategic_engagement_en.pdf

European Commission (2021) *She Figures 2021: Gender in Research and Innovation: Statistics and Indicators*, Directorate-General for Research and Innovation, Publications Office, Available from: https://data.europa.eu/doi/10.2777/06090

Faulkner, W. (2000) 'The power and the pleasure? A research agenda for "making gender stick" to engineers', *Science Technology & Human Values*, 25(1): 87–119, DOI: 10.1177/016224390002500104

Gonsalves, A.J. and Danielsson, A.T. (eds) (2020) *Physics Education and Gender*, Switzerland AG: Springer Nature.

Gvozdanović, J. and Bailey, J. (2021) 'Unconscious bias in academia: a threat to meritocracy and what to do about it', in E. Drew and S. Canavan (eds), *The Gender-Sensitive University: A Contradiction in Terms?*, London: Routledge, pp 110–23, Available from: www.routledge.com/The-Gender-Sensitive-University-A-Contradiction-in-Terms/Drew-Canavan/p/book/978036 7431174?gclid=EAIaIQobChMI-aKintCB9AIVnYBQBh1k3wAmEAAY ASAAEgIlbfD_BwE

Hasse, C. and Trentemöller, A. (2008) *Break the Pattern! A Critical Enquiry into Three Scientific Cultures: Hercules, Caretakers and Worker Bees*, UPGEM-project report, Tartu: Tartu University Press.

HEA (2020) 'Unpublished data obtained from Centre of Excellence for Gender Equality', Dublin: Higher Education Authority.

Heilman, M. and Okimoto, T. (2007) 'Why are women penalized for success at male tasks? The implied communality deficit', *Journal of Applied Psychology*, 92(1): 81–92.

Jammula, D.C. and Mensah, F. (2020) 'Urban college students negotiate their identities to dis/connect with notions of physics', in A. Gonsalves and A. Danielsson (eds) *Physics Education and Gender: Identity as an Analytic Lens for Research*, Cham: Springer, pp 81–96.

LERU (2015) 'Gendered research and innovation: integrating sex and gender analysis into the research process', League of European Research Universities Advice Paper, Leuven, Available from: www.leru.org/files/Gendered-Research-and-Innovation-Full-paper.pdf

LERU (2018) 'Implicit bias in academia: a challenge to the meritocratic principle and to women's careers – and what to do about it', League of European Research Universities Advice Paper, Leuven, Available from: www.leru.org/publications/implicit-bias-in-academia-a-challenge-to-the-meritocratic-principle-and-to-womens-careers-and-what-to-do-about-it

Lundborg, A. and Schönning, K. (2007) *Maskrosfysiker: genusperspektiv på rekrytering, handledning och arbetsmiljö bland Uppsalas fysikdoktorander*, Sweden: Uppsala University, Available from: https://mp.uu.se/docume nts/432512/928760/Maskrosfysiker.pdf/f62d0edd-7311-4f82-9844-ffcd5 3c8124c

Nielsen, M. (2015) 'Make academic job advertisements fair to all', *Nature*, 525(7570): 427, DOI: 10.1038/525427a

Nielsen, M. (2016) 'Limits to meritocracy? Gender in academic recruitment and promotion processes', *Science and Public Policy*, 43(3): 386–99.

Nielsen, M. (2021) 'Gender in academic recruitment and selection', in E. Drew and S. Canavan (eds) *The Gender-Sensitive University: A Contradiction in Terms?*, London: Routledge, pp 28–40, Available from: www.routle dge.com/The-Gender-Sensitive-University-A-Contradiction-in-Terms/ Drew-Canavan/p/book/9780367431174?gclid=EAIaIQobChMI-aKintCB9AIVnYBQBh1k3wAmEAAYASAAEgIlbfD_BwE

Nielsen, M., Bloch, C., and Schiebinger, L. (2018) 'Making diversity work for scientific discovery and Innovation', *Nature Human Behaviour*, 2: 726–34.

Nielsen, M., Alegria, S., Börjeson, L., Etzkowitz, H., Falk-Krzesinski, J., Joshi, A., Leahey, E., Smith-Doerr, L., Woolley, A., and Schiebinger, L. (2017) 'Gender diversity leads to better science', *PNAS*, 114(8): 1740–2, DOI: 10.1073/pnas.1700616114

O'Connor, P. and Irvine, G. (2020) 'Multi-level state interventions and gender equality in higher education institutions: the Irish case', *Administrative Sciences*, 10(98).

Ong, M., Smith, J., and Ko, L. (2017) 'Counter spaces for women of colour in STEM higher education: marginal and central spaces for persistence and success', *Journal of Research in Science Teaching*, 55(2): 206–45.

Sanderson, K. (2021) 'Researchers voice dismay at all-male science Nobels', *Nature*, [online] 8 October, Available from: www.nature.com/articles/d41 586-021-02782-2

Schiebinger, L. (2001) *Has Feminism Changed Science?*, Cambridge, MA: Harvard University Press.

Stewart, A. and Valian, V. (2018) *An Inclusive Academy: Achieving Diversity and Excellence*, Cambridge, MA: MIT Press.

Traweek, S. (1992) *Beamtimes and Lifetimes: The World of High Energy Physicists*, Cambridge, MA: Harvard University Press.

Vainio, A. (2013) 'Beyond research ethics: anonymity as "ontology", "analysis" and "independence"', *Qualitative Research*, 13(6): 685–98.

Wennerås, C. and Wold, A. (1997) 'Nepotism and sexism in peer-review', *Nature*, 387: 341–3.

Wertheim, M. (1997) *Pythagoras' Trousers*, London: Norton.

WIRDEM (2008) *Mapping the Maze: Getting More Women to the Top in Research*, Brussels: European Commission, Available from: http://ec.eur opa.eu/research/science-society/document_library/pdf_06/mapping-the-maze-getting-more-women-to-the-top-in-research_en.pdf

3

Theorizing Gender Inequality in Physics: Gendered Divisions of Labour in the Neoliberal University

Rebecca Lund and Helene Aarseth

Introduction

Physics is routinely described as a discipline characterized by abstract and formative thinking in that it is a scientific pursuit that explicitly seeks to move beyond the boundaries of existing concepts and knowledge (Sakhiyya and Rata, 2019). While formative knowledge production is an important ingredient in all disciplines, physics is arguably a discipline in which the degree of abstract thinking is specifically noteworthy, alongside disciplines such as mathematics and philosophy. Here, the process of formative knowledge creation is understood as a completely creative exercise, involving an experience of deep immersion and absorption in the activity and with the object of study (Fox Keller, 1985; Knorr-Cetina, 1999).

In this chapter, we argue that feminist historical materialism, a theoretical approach that gained traction in the 1980s and 1990s, may be retrieved and revised to shed light on the gendered dynamics and continued gender inequalities and segregation in physics in the contemporary neoliberal governance of academia. The early strands of feminist materialist theories argued that the gendered division of work in capitalist Western societies enforced a deep split between different personality structures and epistemic orientations; a split that meant that men were allowed to be absorbed by the 'abstract labour' of formative knowledge production, while women were seen as immersed in the locally bound and 'concrete labour' of catering for human needs and emotions (Smith, 1974; Hartsock, 1983; Fox Keller,

1985). In the concrete world of attending to human needs and practicalities, everyday life tends to become 'fragmented, episodic, and impossible to plan' (Smith, 1987; Rudberg, 2012). This gendered division of work meant that women scientists would often be 'left with' the invisibilizing tasks of providing connectedness and attending to the bits and pieces that 'proper scientists' did not bother with (see Fox Keller, 1985). Consequently, women provided others, notably men, the conditions for a focused self, enabling them to invest in abstract and formative thinking (Rudberg, 2012; Sakhiyya and Rata, 2019; Federici, 2020). This in part explains why historically men have had better conditions for thriving in the abstract world of the university and with the production of abstract knowledge.

The gendered division of work described by the early feminists has undergone profound changes. In large parts of the Western world, the 'breadwinner and homemaker' society has been replaced with a dual-earner model, leaving the gendered division between care and professional pursuits more blurred (Adkins and Jokinen, 2008). Also, many sectors of the labour market, including higher education, have undergone a so-called 'feminization' referring an increase in female students and academics in previously male-only or male-dominated disciplines (Leathwood and Read, 2008). Moreover, post-industrial work organizations, academia not excepted, place more emphasis on teamwork and networking. As a result, emotional and communicative competence are to an increasing extent considered a requirement in professional life for all employees, regardless of gender (Illouz, 2007; Adkins and Jokinen, 2008). These developments have, to some extent, challenged traditional male-dominated and masculine academic cultures, explicit sexism, and excluding imaginaries of 'the ideal academic' (Lund, 2012).

However, a simultaneous development within higher education has been called a 'remasculinization' largely shaped by a neoliberal restructuring of the academy (Alemán, 2014). This 'remasculinization' operates in subtle and more complex ways. The neoliberal ideology is based on the notion that individuals are equal in the market competition. On the surface, men and women are measured on the exact same performance criteria in the global knowledge economy. Women, and other marginalized groups, have to some extent benefitted from this ideology. By showing proof of productivity, and institutional and economic value – for instance, through publication points or acquiring external funding – they can no longer *explicitly* be questioned as legitimate academics who provide value to the university and discipline (Pereira, 2017). Moreover, gender-equality agendas have explicitly been incorporated into the neoliberal project. A lot of economic and human resources are invested at transnational (for instance, the EU's investment in gender mainstreaming in science and innovation), national, and organizational levels (for instance, the British Athena SWAN programme; the

Norwegian Balance programme), to improve women's abilities to compete on an equal footing with men in academia. These investments have driven *some* positive changes. Still, we argue that the competitive logic spurred by the neoliberal ideology intensifies the underlying conflict between the locally bound attendance to human and emotional needs, and the ability to pursue one's own interests in an uncompromising way. The emphasis on excellence, as defined in the neoliberal academy by competition on standardized metrics, direct people's time, attention, and energies towards the kinds of practices, relationships, and energies that are ascribed value in the global knowledge economy. This instigates a disavowal of the 'concrete labour' of catering for human needs and practicalities. This concrete labour, such as paying attention to a student who struggles with her thesis, is typically not directly measurable, and so excluded from the standarized measures. When academic success is preconditioned on the capability to disavow this labour, the split between care and competition analysed by the early feminist contributions is intensified. Reconceptualized in a manner that takes the increased complexities of the relationship between capitalism, gender, and work into account, these contributions may help us to understand why and how the new governance through excellence undermines continuous efforts to achieve gender balance.

The chapter is organized as follows: first, we provide some insight into neoliberalism and the marketization of the academy, which is both significant for our identification of *why* gender inequality is persistent in sciences and physics, and for positioning our suggested theorization of how gender inequality is (re)produced. Second, we provide a literature review of some key contributions to understanding women's position in science and locate ourselves in relation to these. Third, we turn to retrieving feminist historical materialism, specifically for explaining split gendered energies, and how these are produced in the context of the neoliberal academy and externally defined standards of quality. We finalize the chapter by concluding how our suggested theoretical approach can help explain persistent gender inequality in physics, while also pointing to the implications this mode of theorizing has for how we may counter inequality.

Setting the scene: neoliberalism and the global reorganization of science

Neoliberalism is a political ideology and set of policy practices that extends market logics to sectors previously governed by the state, based on the rationale that state governance is authoritarian and inefficient (Foucault, 2008). The neoliberal state actively 'constructs and protects markets' based on the logic that competition not only leads to more cost-efficient and higher quality services, but also provides choice and thereby individual freedom (Lave et al, 2010: 661). Neoliberal ideology became highly influential from the 1970s and 1980s in the UK, Margaret Thatcher's and, in the US, Ronald Reagan's governments, and also became the dominant philosophy of transnational organizations, such as the Organization for Economic Corporation and Development (OECD), the International Monetary Fund (IMF), and the World Trade Organization (WTO). From then on, these ideas have achieved status as inevitable 'normative frameworks' for how to handle all sorts of social and economic problems (Moore et al, 2011: 512). The 'effects of globalized neoliberalism are unevenly distributed' across countries and institutions (Moore et al, 2011: 208) depending on particular socio-historical conditions and degrees of resistance. Nonetheless, it is possible to identify 'family resemblances' across specific contexts: '(1) markets are preferred over government as policy instrument; (2) trade liberation is preferred above protectionism; and (3) social problems and inequities are approached through a logic of 'individual responsibilization', decentralization, and economic development for increasing the standard of living, rather than through state-governed economic redistribution' (Moore et al, 2011: 208).

During the 1990s and 2000s, national science policies were remolded through market-based solutions and universities were given a new role. They were increasingly perceived as a central actor in global competition and as drivers of a new knowledge economy (Slaughter and Leslie, 1997; Wright and Shore, 2017). A key element in the neoliberalization of universities was performance of academic units based on numerical standards. Scholars from the social sciences and humanities have unpacked the effects of an 'audit culture' describing how academics are increasingly made 'objects of managerial discipline' and 'illiberal governance', and a resulting loss of departmental autonomy (Shore, 2008). The changing role of universities as actors in global competition has shifted 'prestige-relations' *between* and *within* universities and disciplines (Moore et al, 2011; Willmott, 1995). Prestige at organizational, disciplinary, and individual levels is increasingly dependent on productivity, the acquisition of funding, industrial innovation, or societal impact, in a manner that, arguably, negatively affects the perceived value of basic sciences and formative knowledge creation (Sakhiyya and Rata, 2019). While this chapter does not offer an empirical analysis of these processes, it offers theoretical tools for exploring the gendered effects of neoliberalism on physics and sciences more generally. Our point of entry is that physics as a basic science, like other disciplines, is being shaped by a culture of commerce. A characteristic feature of the culture of commerce is the valuation of flexible labour. In a research and prestige economy increasingly defined by external funding, this makes sense. But from the perspective of those in the position of continually having to secure the next temporary contract and perhaps in time qualify for a permanent contract, it results in performance pressure and emotional pressure: you continually have to prove your value on standardized metrics of excellence. This individualistic competitive system responsibilizes people for their success or failure (Gershon, 2011; Shore, Wright and Però, 2011) in ways that have gendered consequences (Thun, 2020; Holter and Snickare, 2022).

Background: women and science

A vast amount of research has been dedicated to explore the gendered divisions of labour at home and at university since the 1960s. While both men and women find juggling family life and working life difficult, women academics find this more challenging. For instance, Elizabeth O'Laughlin and Lisa Bischoff (2005), in their study of dual-academic-career two-parent families, found that women feel more pressured to fulfil their twin roles as parents and academics. Women's stress related to these conflicting roles resulted in low self-esteem, more instances of depression, decreased job satisfaction, and less satisfaction with one's parental role. This not only results in lower quality and quantity in work performance, but also ultimately leads to more women opting out of academia because they are denied tenure or other promotions (O'Laughlin and Bischoff, 2005; see also Rafnsdottir and Heijstra, 2011).

Along these lines, EU research reports have shown that women in heterosexual relationships are more likely than men to take parental leave and part-time positions in order to ensure a sustainable work–life balance. As outlined in Chapters 4 and 6 of this book, a recent study of physics doctoral students in Israel reports similar findings and points to the extra burden for mothers in heterosexual relationships. This is also the case in Nordic countries with relatively favourable paternity leave policies and other 'family-friendly' policies (see Brandth and Kvande, 2018). This has been shown to have repercussions for women's salary levels, pensions, promotion, and tenure. It has been suggested that this pattern would change if men were more involved in care work, and if historical and cultural practices resulting in the devaluing of women's work, tasks, and skills were confronted (Duvander et al, 2005; European Commission, 2013).

Closely related to wage levels and academic merit is the question of gendered success rates in achieving competitive research grants, showing that women principal investigators in the natural sciences have significantly lower success rates than men (European Commission, 2009; Husu and Callerstig, 2018; Steinþórsdóttir et al, 2020) – something which also, of course, relates to the fact that larger funding resources are allocated to research fields of research dominated by men (Henningsen and Liestøl, 2013; Thun, 2020; Holter and Snickare, 2022). Research by Etzkowitz et al (2000) has in addition shown that women, compared with men, have fewer collaborative relations beyond their immediate department, and that this negatively impacts publication productivity and success. Associated with this, women were found to be less mobile and thereby less internationally visible. Compared with men, they do not participate as frequently in international conferences, conduct visiting scholarships abroad, or receive invitations to guest lecturing. This can be explained by women taking personal and family attachments more seriously than men (Etzkowitz et al, 2000), and that women take on heavier teaching loads (She Figures, 2012) and care responsibilities at work. In practice this means that women are more tied to their home institutions and have less time for travel- and research-related activities, for example (Van den Brink and Benschop, 2012).

While the previously mentioned studies offer important insights, they also display limited (if any) consideration of particular historical, social, and economic processes shaping the conditions of women and men's work in contemporary academia. Therefore, they often end up pointing towards individualized solutions to the problem of gender inequality. One example of this is mentoring schemes that encourage women to prioritize the kinds of activities that are considered valuable (Rottenberg, 2014; Magnussen et al, 2022). The aim is to ensure that women make better cost–benefit calculations and feel more comfortable with 'leaning in' (Sandberg, 2013; Lund, 2020).

Another line of studies that contribute to our current understanding of gender in contemporary academia are those that theorize work, organizations, and institutions, such as universities and higher education, as shaped in capitalist and patriarchal power relations. Socialist feminist theorizing includes the work of Joan Acker (1990, 2006). Acker argued that organizations and institutions are neither gender-neutral or asexual, and suggested that we study 'how gender provides the subtext for arrangements of subordination', especially those that organize work relations (Acker, 1990: 155). Acker directed attention to features of organizations that constitute gender and inequality arguing that organizations are significant settings for the reproduction of capitalist and patriarchal societal and economic orders. The very notion of 'work' and what Acker called 'the ideal worker' are based on masculine, middle-class, heterosexual, and White assumptions vested in a gendered division of labour. The physics discipline is historically a highly competitive and male-dominated terrain (Eran Jona and Nir, 2021); membership of the community depends on high levels of commitment and the notions of the ideal scientist as 'an individual great man' is deeply rooted in the culture (see Brage and Drew, Chapter 2 in this book).

Since Acker's early theorization, numerous empirical studies have explored the constraints faced by women in working life and unpacked the ways in which gender inequalities in status and material circumstances are reproduced (for example, Van den Brink and Benschop, 2012). While progress has been made to overcome many of the most obvious structural, institutional, cultural disadvantages and discrimination practices, gendered hierarchies continue to exist in more or less subtle forms that remain taken for granted and hard to challenge (Thun, 2020; Clavero and Galligan, 2021; Eslen-Ziya and Yildirim, 2022).

Recent developments, shaped by neoliberal reforms of higher education, indicate a remasculinization of academia, not least driven by the link forged between the ideal neoliberal responsibilized subject and increasingly the narrow conceptions of what constitutes merit and valuable academic work within the academic prestige and research economy, for example (Lund, 2012; Davies and O'Callaghan, 2014; Jenkins, 2014). Competition for funding and positions are increasingly organized according to standardized measures such as number of high-impact journal articles and volume of funding (Orupabo and Mangset, 2021). While this would seem to imply that women and men are treated on equal terms, the 'meritocratic terms' are still rooted in a fundamentally masculine ideal (Thun, 2020). This is an autonomous and competitive individual who passionately invests in 'what counts' (Lund and Tienari, 2019), leaving the non-meriting tasks, such as teaching and supervision, staff meetings, and academic housework to others (Babcock et al, 2017; Guarino and Borden, 2017). Arguably, the new standardized measures come with a shift from one prestige system based on expert-based power, to another system based on market-based power anchored in 'academic capitalism' (Metcalfe and Slaughter, 2008). What is valorized here, is the competitive, individualistic, and entrepreneurial individual, whereas individuals invested in care work and community both outside and inside academia are disadvantaged (Alemán, 2014; Eran Jona and Nir, 2021).

Socialist feminist analysis has traditionally worked from a distinction between value-producing activities inside and outside the formal labour processes. This has included a distinction between those value-producing activities that gain institutional representation and rewards (for example, in the form of employment, promotion, salary) and those value-producing activities that are invisible in institutional representations and are not rewarded. The invisibilized everyday activities of some people form a taken-for-granted subtext and prerequisite for other people to approximate neoliberalized conceptions of the ideal academic (Acker, 2006; Lund, 2012). Feminine reproductive emotional, care, and affective work in the home and workplace are not comparable and measurable along lines of 'masculine' abstract units of time and tasks.

Recent developments in capitalism involve ascribing value to the so-called 'feminine' emotional and 'non-quantifiable' activities and have included them in the formal labour process (Adkins and Jokinen, 2008). Practices previously labelled feminine have become an asset on which capitalist accumulation depends (Adkins, 2008; Eisenstein, 2017). This has made some scholars question whether socialist feminist theories are still relevant. Indeed, the development would seem to challenge the gendered division of labour that constitutes women as women and men as men in socialist feminist theorizing, as both women and men, in order to be considered valuable, must perform 'feminine' and emotional work. Central to this way of theorizing gender inequality, is that femininity and masculinity may be done by both men and women. The fact that political discourses may favour advancement of women and 'feminine values' in the labour market does not necessarily result in gender equality within organizations. Although renewed emphasis on emotional and competitive competence may challenge gender stereotypes, it is perfectly possible that this reappraisal coexist with 're-traditionalization' of gender hierarchies. For instance, so-called feminine practices linked to the catering of human needs and relations may be rewarded when men do them, but be considered 'natural' when women do them (Adkins, 2004; Blackmore, 2014; Lund et al, 2019).

While these theories explain how gender inequality is shaped in wider processes of capitalism and patriarchy, thus turning our attention away from individual responsibilities and towards a critique of social structures, we suggest that the structuring of gender inequality has another layer which is not included in the above. More specifically, the layer accounting for how capitalism enforces a split in gendered subjectivities and epistemic orientations in terms of abstract versus concrete labour (Hartsock, 1983; Fox Keller, 1985; Smith, 1987). In spite of less explicit gender divisions, the split between those who provide the conditions that enable others to acquire the mental space necessary for engaging in complex abstract thinking may actually be reproduced. This split, we argue, becomes increasingly relevant in the context of neoliberal restructuring that enforces people to channelling of time and energies towards a specific and narrowly defined outcome, to 'doing what counts'.

To explain this split, we suggest reinvigorating early strands of historical materialist feminism. We draw on their key argument that the gendered division of work in capitalist Western society enforces a deep split between different subjectivities and epistemic orientations (Hartsock, 1983; Fox Keller, 1985; Smith, 1987; Fraser and Jael, 2018; Federici, 2020). We suggest a reading that invokes how divisions of labour in the context of the neoliberal academy produces a splitting in the directedness of energies on the one hand, and, on the other, also suggest that this division and splitting is driven by textually mediated discourses of excellence and standards of quality.

Retrieving historical materialist conceptions of the gendered division of labour

The directedness of energies

While the gendered division between reproductive and productive labour is less clear-cut in contemporary neoliberal and 'feminized' post-industrial societies, the gendered subtext originally theorized by Acker (Acker, 1990, 2006) has not disappeared, but rather intensified in the university. This intensification, we argue, is driven by increasingly narrow notions of focusing on 'what counts' and how that shapes modes of relating to each other.

Early historical materialist feminist contributions emphasized the relational and 'need orientated' disposition linked to the work of mothering and domestic work, for example (Hartsock, 1983; Smith, 1987). The need-orientated mode of being and relating was in turn linked to gendered splits in socialization, theorized as originating in pre-linguistic processes of separation and individuation in the early infancy (Chodorow, 1978; Hartsock, 1983; Fox Keller, 1985). The psychoanalytic notion of gendered identities resulted in a forceful critique for being essentializing and deterministic. However, freed from this notion of gendered identities, we argue that this strand of historical materialism may still offer productive tools for conceptualizing a split in modes of directing energies that make up the gendered subtext in contemporary neoliberal universities (see, for example, Fraser and Jaeggi, 2018; Federici, 2020). Feminist theorists have, as showed earlier, emphasized how this splitting of energies is gendered. Both women and men encounter this tension and choice, but research shows that women invest more in caring relations (Eran Jona and Nir, 2021; Holter and Snickare, 2022). This means that people with caring responsibilities are engaged in a 'continuous struggle on the boundaries of "poverty" in terms of their possibilities to operate in society as self-assured and self-evidently worthy people exerting their capacities effectively and legitimately' (Jonasdottír, 1994: 225; see also Gunnarsson, 2016). Arguably, one could claim that the structural organization of neoliberal societies can be perceived as reproduced through a self-sustaining and self-energizing split between competitive and relational investments. This splitting of energies is not grounded in a gendered split between an essentially competitive masculinity and relational femininity but, rather, is incited by a gendered division of work and the societal mode of organizing practices that emanates from it. The point that we take with us from the work of these early materialist feminists is this: in order to produce *and* preserve human life-energies, some modes of production must work according to a logic that differs from that of 'return on investment' (Jonasdottír, 1994). This argument is based in a foundational assumption, that human thriving and surviving requires being made subject to loving energies, care, and ecstasy. The receiving of such energies constrains and enables human life possibilities and unfolding (Gunnarsson, 2016).

Feminist Marxist philosopher Anna Jonasdottír developed the concept of *love power* to describe modes of production aimed at sustaining and nurturing other people's energies and thus providing them 'life-enhancing opportunities' (Gunnarsson, 2016). Through theorizing gendered divisions of love power, Jonasdottír articulates a gendered subtext of exploitation, where men extract more love power from women than they give back (Jonasdottír, 1994, 2013).

The gendered division of work increasingly plays out at the level of affects. In contemporary neoliberal universities, containment of anxiety becomes increasingly difficult (Gill, 2009; O'Neill, 2014). Instrumentalization and competition diminish what Young (1994) describes as the 'mental space' where human vulnerabilities, anxieties, and longings can be contained, reflected upon, and savoured. Psychosocial theorists have described how marketization and audit culture fuels a range of defence mechanisms, including the projection of vulnerability and neediness onto others (Layton, 2020). In fact, the neoliberal competitive culture may be interpreted as a defence mechanism in itself! The obsession with labelling, measuring, and ranking can be seen as defensive efforts aimed at solidifying a frail identity. This defence produces 'a distancing from the actual person, who instead becomes synonymous with standardized task performance' (O'Neill, 2014: 7). The lack of 'mental space' induces a self-energizing dynamic where the need to handle anxieties through positional competition diminishes the possibility for intra-psychic dialogue, in turn fuelling the need for mastery through positioning, and so on. Therefore, as Layton (2020) has argued, societal formations characterized by intensive competitiveness tend to produce environments where it becomes essential to *get rid* of feelings of dependency and vulnerability, whatever the cost. In order to optimize their competitive capacities, the selves required in market competition must be able to 'hold themselves together' and effectively manage the world around them. This requires a distancing towards one's own as well as other people's needs and vulnerabilities.

An organization of academic work that emphasizes the furthering of the subject's competitive advantage channels human energies in a way that necessitates the exploitation of other human needs and potentials. This, in turn, can fuel modes of being and relating founded on self-assuredness and the projection of vulnerability and weakness onto others (Gunnarsson, 2016); that is, the seeds of a toxic individualized and competitive culture. In the analysis of Rebecca Lund and Janne Tienari (2019), libidinal strivings are channelled into an individualist passion detached from the concrete concerns of everyday life, such as catering to human needs, relations, and emotions in one's immediate surroundings. There are increasing indications that this includes a gendered division of work in the academic organization, leaving women academics with more responsibility for relational and domestic functions linked to supervising, teaching, and administration (Lund, 2015; Babcock et al, 2017; Lund and Tienari, 2019). Feminist critics have noted how the requirements for self-assertiveness and 'sequestrated passion' that fuel the ability to marketize oneself and one's 'outputs' are typically linked to the performance of masculinity (Lund and Tienari, 2019). Subjects who are less forced to be in contact with their own vulnerabilities are inclined to boast and are less susceptible to self-blaming (Baker and Brewis, 2020) and toxic shame (Gill, 2009). This dynamic will further deepen splits between life-sustaining energies that cater for human needs and relations on the one hand, and investments in competitive outcomes on the other (Aarseth, 2022). Arguably, this split does not only, or perhaps even primarily, foster a devaluation of women, but also is linked to a more profound alienated love power, producing a masculinized neoliberal subject fully dependent on, yet in denial of, vulnerability, connectedness, and care (Gunnarsson, 2016).

The directedness of energies is one level on which neoliberalism shapes the gendered subtext of academic working life. Another mode is that of discourses, which we shall turn to now.

Discursively enhanced splits

The feminization of capitalism (Hochschild, 1983; Standing, 1999; Adkins and Jokinen, 2008) means that men, in order to be considered competent in contemporary working life, are also expected to invest in relations to people. However, the *kinds of relations* may differ from the ones women invest in (Skeggs, 2004). This in part explains why gender inequality and segregation remains a significant feature of academic working life today, despite political and organizational investments in gender equality, neoliberalism, and numerical standards of excellence (Lund, 2015). Another layer of theorizing the (re)production of split energies under neoliberalism is by pointing towards the textual standards shaping academic practices, and the discourses emanating from these. Indeed, the neoliberal university is thoroughly textually mediated and governed (Shore and Wright, 2017). These discourses discourage and make invisible investments in the local, while encouraging an orientation towards and investment in the global: objectified texts 'coordinate(s) the acts, decisions, policies, and plans of actual subjects as the acts, decisions, policies and plans of large-scale organizations' (Smith, 1990: 61). In other words, the institutional discourses mediated in organizational texts shape actual social practices of knowledge production and produce specific gendered organization within academia. While the texts associated with policy reform and strategic management may have the appearance of neutrality, the gendered subtexts are 'revealed' when local actualities and experiences point to the ways in which boundaries are drawn and categories of distinction are created in the process of their local enactment (Smith, 1990: 65).

As pinpointed earlier, neoliberal policy and practice involves a responsibilization of subjects for their success or failure. In order to be successful in the competition for a position, subjects must invest in the kinds of practices, relationships, and energies that are ascribed value in the global knowledge economy, doing what counts in accordance to standards increasingly set out in texts, such as the Research Excellence Framework in the UK (Wright, 2014; Wright and Shore, 2017). This involves, at a very concrete level, allowing oneself to be carried away from home (Parker and Weik, 2014). Martin Parker and Elke Weik argue that 'the original notion of intellectual detachment and academic freedom has developed into a demand for social and moral detachment by the ever-growing circuit of international "visibility" as celebrated at international conferences'. This 'excludes all those whose attachment to persons or causes requires bodily presence, and such an exclusion transforms the contents and values of academic knowledge' (Parker and Weik, 2014: 167). Of course, neoliberalism cannot alter natural laws which are the object of exploration in physics. Yet, it can alter the conditions of science by shaping how value and importance are attached to specific kinds of experiments, collaborations, and funding. It can also alter the extent to which people can accomplish the kind of 'mental space' imperative in the pursuit of complex, formative thinking (Young, 1994: Sakhiyya and Rata, 2019; Aarseth, 2022).

Rebecca Lund and Janne Tienari (2019) showed how 'potential', a futurity-oriented evaluation of competence, becomes gendered through discourses of care and passion, generated by the neoliberal culture of commerce. In order for some people to detach themselves from activities that do not count and orient themselves towards competing in the global knowledge economy, others must do the work of maintaining the local, invisibilized, yet detrimental, for the survival of the university. Expanding on Parker and Weik's (2014) analysis, Lund and Tienari (2019: 98) argue that 'material and epistemic detachment from the local is framed as potential and how masculinized passion directs academics to do what counts, while feminized and locally bound care is institutionally appreciated only as far as it support individualized passion'.

Conclusion

The turn to neoliberalism, marketized competition, and standardized measures of quality has to some extent countered gender bias in competitive processes, and in fact has been explicitly presented as a mode of countering homosociality and the maintenance of male hegemony (Bird, 1996) through transparent measures, recruitment, and promotion processes (Lund, 2015). Despite these measures it seems, however, that gender inequality continues to be a problem. To explain this, we have suggested that there is a need to pay attention to the effects of intensified market competition and how it shapes practices and subjective orientations. To do so we have proposed that retrieving early historical materialist feminist ideas enable enhanced insights into these mechanisms.

The early historical materialist strands were particularly focused on explaining how women's reproductive labour was exploited in ways that reproduced a binary division between men and women as two distinct groups with distinct psychosocial personality structures. While this conception appears as problematic and less apt to account for current gender dynamics, we have shown that the ideas underpinning these conceptualizations – specifically, gendered splits in the directedness of energies – provide conceptual tools particularly fruitful for deepening our understanding of persistent gender inequality in physics.

We have pointed at two levels of analysis that arise from this insight and enhance our understanding of the gendered subtexts of neoliberalism: first, a deepened split between competitive power founded on a disavowal of vulnerability and dependence on the one hand, and a detriment other-orientation partly defined by the need to stitch together the residuals of the 'false' independence of the other; and second, a textually mediated enhancing of these splits resulting from marketization and the downplaying or invisibilization of work that does not count in accordance with externally established standards.

Consequently, we argue that what appears to be forms of gender-neutral- and gender-balance-promoting forms of governance actually spurs processes of increased splitting between reproductive and productive practices in ways that not only preserve, but also potentially deepen the current gender gap. The structural conditions of academic work and employment furthered by neoliberal reform and marketization exacerbate historically rooted individualistic tendencies, gendered disciplinary hierarchies, and a masculine performance culture within physics that systematically devalues reproductive labour and people with caring responsibilities. The implication of this insight is that, in order to challenge the gender gap, we must challenge this underlying split. We must provide space for 'the mental space' required in the kinds of abstract, formative thinking that are crucial within physics in a way that does not depend upon the exclusion and marginalization of the emotional investment in care outside and inside the university.

References

Aarseth, H. (2022) 'The implicit epistemology of metric governance. New conceptions of motivational tensions in the corporate university', *Critical Studies in Education*, 63(5): 589–605.

Acker, J. (1990) 'Hierarchies, jobs, bodies: a theory of gendered organizations', *Gender and Society*, 4(2): 139–58.

Acker, J. (2006) 'Inequality regimes: gender, class, and race in organizations', *Gender and society*, 20(4): 441–64.

Adkins, L. (2004) 'Reflexivity: freedom or habit of gender?', *The Sociological Review*, 52(2): 191–201.

Adkins, L. (2008) 'From retroactivation to futurity: the end of the sexual contract?', *NORA—Nordic Journal of Feminist and Gender Research*, 16(3): 182–201.

Adkins, L. and Jokinen, E. (2008) 'Introduction: gender, living and labour in the fourth shift', *NORA—Nordic Journal of Feminist and Gender Research*, 16(3): 138–49.

Alemán, A.M.M. (2014) 'Managerialism as the "new" discursive masculinity in the university', *Feminist Formations*, 107–34.

Babcock, L., Recalde, M.P., Vesterlund, L., and Weingart, L. (2017) 'Gender differences in accepting and receiving requests for tasks with low promotability', *American Economic Review*, 107(3): 714–47.

Baker, D. and Brewis, D. (2020) 'The melancholic subject: a study of self-blame as a gendered and neoliberal response to the loss of the "perfect worker"', *Accounting, Organizations and Society*, 82: 101093.

Bird, S.R. (1996) 'Welcome to the men's club: homosociality and the maintenance of hegemonic masculinity', *Gender and Society*, 10(2): 120–32.

Blackmore, J. (2014) 'Disciplining academic women: gender restructuring and the labour of research in entrepreneurial universities', in M. Thornton (ed) *Through a Glass Darkly: The Social Sciences Look at the Neoliberal University*, Canberra: Australian National University Press, pp 179–94.

Brandth, B. and Kvande, E. (2018) 'Masculinity and fathering alone during parental leave', *Men and Masculinities*, 21(1): 72–90.

Chodorow, N. (1978) *The Reproduction of Mothering: Psychoanalysis and the Sociology of Gender*, Berkeley, CA: University of California Press.

Clavero, S. and Galligan, Y. (2021) 'Delivering gender justice in academia through equality plans? Normative and practical challenges', *Gender, Work and Organization*, 28(3): 1115–32.

Davies, H. and O'Callaghan, C. (2014) 'All in this together? Feminisms, academia, austerity', *Journal of Gender Studies*, 23(3): 227–32.

Duvander, A.Z., Ferrarini, T., and Thalberg, S. (2005) *Swedish Parental Leave and Gender Equality: Achievements and Reform Challenges in a European Perspective*, Stockholm: Institute for Futures Studies, p 11.

Eisenstein, H. (2017) 'Hegemonic feminism, neoliberalism and womenomics: empowerment instead of liberation?', *New Formations*, 91: 35–49.

Eran Jona, M. and Nir, Y. (2021) 'Choosing physics within a gendered power structure: the academic career in physics as a "deal"', *Physical Review, Physical Education Research*, 17(2), DOI: 10.1103/PhysRevPhysEducRes.17.020101

Eslen-Ziya, H. and Yildirim, T.M. (2022) 'Perceptions of gender challenges in academia', *Gender, Work and Organization*, 29(1): 301–8.

Etzkowitz, H., Kemelgor, C., and Uzzi, B. (2000) *Athena Unbound: The Advancement of Women in Science and Technology*, Cambridge: Cambridge University Press.

European Commission (2009) *The Gender Challenge in Research Funding*, Brussels: European Union.

European Commission (2013) *The Role of Men in Gender Equality – European Strategies and Insights*, Brussels: European Union.

Federici, S. (2020) *Beyond the Periphery of the Skin: Rethinking, Remaking, and Reclaiming the Body in Contemporary Capitalism*, Oakland, CA: PM Press.

Foucault, M. (2008) *The Birth of Biopolitics: Lectures at the Collège de France: 1978–1979*, New York: Springer.

Fox Keller, E. (1985) *Reflections on Gender and Science*, New Haven, CT: Yale University Press.

Fraser, N. and Jaeggi, R. (2018) *Capitalism: A Conversation in Critical Theory*, Cambridge: Polity.

Gershon, I. (2011) 'Neoliberal agency', *Current Anthropology*, 52(4): 537–55.

Gill, R. (2009) 'Breaking the silence: the hidden injuries of neo-liberal academia', in R. Flood and R. Gill (eds) *Secrecy and Silence in the Research Process: Feminist Reflections*, London: Routledge, pp 228–44.

Guarino, C.M. and Borden, V.M.H. (2017) 'Faculty service loads and gender: are women taking care of the academic family?', *Research in Higher Education*, 58(6): 672–94.

Gunnarsson, L. (2016) 'The dominant and its constitutive other: feminist theorizations of love, power and gendered selves', *Journal of Critical Realism*, 15(1): 1–20.

Hartsock, N.C. (1983) 'The feminist standpoint: developing the ground for a specifically feminist historical materialism', in S. Harding, and M.B. Hintikka (eds) *Discovering Reality*, New York: Springer, pp 283–310.

Henningsen, I. and Liestøl, K. (2013) '*Likestilling i akademia–Er eksellense for menn og Grand Challenges for kvinner?*', *Tidsskrift for kjønnsforskning*, 37(3–4): 348–61.

Hochschild, A.R. (1983) *The Managed Heart*, Berkeley, CA: Berkeley University Press.

Holter, Ø. and Snickare, L. (2022) *Gender Equality in Academia – From Knowledge to Change*, Oslo: Cappelen Damm.

Husu, L. and Callerstig, A.C. (2018) *Riksbankens Jubileumsfonds beredningsprocesser ur ett jämställdhetsperspektiv*, Stockholm: Riksbankens Jubileumsfond.

Illouz, E. (2007) *Cold Intimacies: The Making of Emotional Capitalism*, London: Polity Press.

Jenkins, E.K. (2014) 'The politics of knowledge: implications for understanding and addressing mental health and illness', *Nursing Inquiry*, 21(1): 3–10.

Jonasdottír, A.G. (1994) *Why Women Are Oppressed*, Philadelphia, PE: Temple University Press.

Jonasdottír, A.G. (2013) 'Love studies: a (re)new(ed) field of feminist knowledge interests', in A.G. Jonasdottír and A. Ferguson (eds) *Love: A Question for Feminism in the 21st century*, London: Routledge, pp 11–33.

Knorr-Cetina, K. (1999) *Epistemic Cultures: How the Sciences Make Knowledge*, Cambridge, MA: Harvard University Press.

Layton, L. (2020) *Toward a Social Psychoanalysis: Culture, Character, and Normative Unconscious Processes*, London: Routledge.

Leathwood, C. and Read, B. (2008) *Gender and the Changing Face of Higher Education: A Feminized Future*, London: Open University Press.

Lund, R. (2012) 'Publishing to become an "ideal academic": an institutional ethnography and a feminist critique', *Scandinavian Journal of Management*, 28(3): 218–28.

Lund, R. (2015) *Doing the Ideal Academic: Gender, Excellence and Changing Academia*, Helsinki: Aalto University.

Lund, R. (2020) 'Becoming a professor requires saying "no!": merging gender equality and quality agendas in a Norwegian gender balance project', in H.L. Smith, C. Henry and H. Etzkowitz (eds) *Gender, Science and Innovation*, London: Edward Elgar, pp 35–57.

Lund, R. and Tienari, J. (2019) 'Passion, care, and eros in the gendered neoliberal university', *Organization*, 26(1): 98–121.

Lund, R., Meriläinen, S., and Tienari, J. (2019) 'New masculinities in universities?', *Gender, Work and Organization*, 26(10): 1376–97.

Magnussen, M.-L., Lund, R., and Mydland, T. (2022) 'Uniformity dressed as diversity? Reorienting female associate professors', in G. Griffin (ed) *Gender Inequalities in Tech-Driven Research and Innovation: Living the Contradiction*, Bristol: Policy Press, pp 109–23.

Metcalfe, A. and Slaughter, S. (2008) 'The differential effects of academic capitalism on women in the academy', in J. Glazer-Raymo (ed) *Unfinished Agendas: New and Continuing Gender Challenges in Higher Education*, Baltimore, MD: Johns Hopkins University Press, pp 80–111.

Moore, K., Kleinman, D.L., Hess, D., and Frickel, S. (2011) 'Science and neoliberal globalization: a political sociological approach', *Theory and Society*, 40(5): 505–32.

O'Laughlin, E.M. and Bischoff, L.G. (2005) 'Balancing parenthood and academia: work/family stress as influenced by gender and tenure status', *Journal of Family Issues*, 26(1): 79–106.

O'Neill, M. (2014) 'The slow university: work, time and well-being', *Forum: Qualitative Social Research*, 15(3).

Orupabo, J. and Mangset, M. (2021) 'Promoting diversity but striving for excellence: opening the "Black Box" of academic hiring', *Sociology*, DOI: 00380385211028064

Parker, M. and Weik, E. (2014) 'Free spirits? The academic on the aeroplane', *Management Learning*, 45(2): 167–81.

Pereira, M.D.M. (2017) *Power, Knowledge and Feminist Scholarship: An Ethnography of Academia*, London: Taylor and Francis.

Rafnsdottir, G.L. and Heijstra, T.M. (2011) 'Balancing work–family life in academia: the power of time', *Gender, Work and Organization*, 20(3): 283–96.

Rottenberg, C. (2014) 'The rise of neoliberal feminism', *Cultural Studies*, 28(3): 418–37.

Rudberg, M. (2012) 'Gender, knowledge and desire', in K. Jensen, L.C. Lahn, and M. Nerland (eds) *Professional Learning in the Knowledge Society*, Rotterdam: Sense Publishers, pp 179–93.

Sakhiyya, Z. and Rata, E. (2019) 'From "priceless" to "priced": the value of knowledge in higher education', *Globalisation, Societies and Education*, 17(3): 285–95.

Sandberg, S. (2013) *Lean In: Women, Work and the Will to Lead*, New York: Knopf.

She Figures (2012) *Gender in Research and Innovation*, Brussels: European Commission.

Shore, C. (2008) 'Audit culture and illiberal governance: universities and the politics of accountability', *Anthropological Theory*, 8(3): 278–98.

Shore, C., Wright, S., and Però, D. (eds) (2011) *Policy Worlds: Anthropology and the Analysis of Contemporary Power* (vol 14), New York: Berghahn Books.

Shore, C. and Wright, S. (eds) (2017) *The Death of the Public University? Uncertain Futures for Higher Education in the Knowledge Economy*, New York: Berghahn Books.

Skeggs, B. (2004) 'Exchange, value and affect: Bourdieu and "the self"', *Sociological Review*, 52(2): 75–95.

Slaughter, S. and Leslie, L.L. (1997) *Academic Capitalism: Politics, Policies, and the Entrepreneurial University*, Baltimore, MD: John Hopkins University Press.

Smith, D.E. (1974) 'Women's perspective as a radical critique of sociology', *Sociological Inquiry*, 44(1): 7–13.

Smith, D.E. (1987) *The Everyday World as Problematic: A Feminist Sociology*, Toronto: University of Toronto Press.

Smith, D.E. (1990) *The Conceptual Practices of Power: A Feminist Sociology of Knowledge*, Toronto: University of Toronto Press.

Standing, G. (1999) 'Global feminization through flexible labor: a theme revisited', *World Development*, 27(3): 583–602.

Steinþórsdóttir, F.S., Einarsdóttir, Þ., Pétursdóttir, G.M., and Himmelweit, S. (2020) 'Gendered inequalities in competitive grant funding: an overlooked dimension of gendered power relations in academia', *Higher Education Research and Development*, 39(2): 362–75.

Thun, C. (2020) 'Excellent and gender equal? Academic motherhood and gender blindness in Norwegian academia', *Gender, Work and Organization*, 27(2): 166–80.

Willmott, H. (1995) 'Managing the academics: commodification and control in the development of university education in the UK', *Human Relations*, 48(9): 993–1027.

Wright, S. (2014) 'Knowledge that counts: points systems and the governance of Danish universities', in A.I. Griffith and D.E. Smith (eds) *Under New Public Management: Institutional Ethnographies of Changing Front-Line Work*, Toronto: University of Toronto Press, pp 294–337.

Van den Brink, M. and Benschop, Y. (2012) 'Slaying the seven-headed dragon: the quest for gender change in academia', *Gender, Work and Organization*, 19(1): 71–92.

Wright, S. and Shore, C. (eds) (2017) *Death of the Public University?: Uncertain Futures for Higher Education in the Knowledge Economy* (Vol 3), New York: Berghahn Books.

Young, R.M. (1994) *Mental Space*, London: Process Press.

PART II

Researching Gender and Physics

4

The Academic Career in Physics as a 'Deal': Choosing Physics within a Gendered Power Structure

Meytal Eran Jona and Yosef Nir

Introduction

The small proportion of women among university students and academic staff in maths-intensive disciplines is an issue of much concern across the globe. Over the last two decades, much research has been carried out in this regard, and effort invested in trying to improve the situation. Of particular interest is the field of physics, where the gender imbalance in the academy is particularly severe, and where women's participation has shown no significant increase in spite of dedicated efforts in the US (Porter and Ivie, 2019), in Europe (Blue and Cid, 2018), and globally (Cochran et al, 2019).

In this chapter, we discuss a research study, which focuses on the gender imbalance in the field of physics in the Israeli academy. The percentages of women among physics Israeli students and academic staff are even poorer than in the US and Western Europe: 16 per cent among graduate students and 6 per cent among faculty members (Eran Jona and Nir, 2019) compared with close to 20 per cent in various Western democracies. These low rates are even more striking when compared with other fields in the Israeli academy, such as medicine, where, in 2018, women constitute 69 per cent among graduate students and 35 per cent among faculty members, or biology, where the proportion is 58 per cent and 30 per cent, respectively.

Looking at the data, we asked ourselves why so few women pursue an academic career in physics. In Chapter 6, we will focus on the difficulties faced by men and women PhD physics students and highlight the extra hurdles that women students encounter. Another main career stage that contributes to the gender imbalance among academic staff in physics is the transition from PhD to postdoc. At this stage, many women students give up on an academic career path (Gofen, 2011). In this chapter, we explore the causes of the gender imbalance at this stage, by focusing on the decision-making junction between PhD and postdoc.

In this context, we pose the following three research questions: How do PhD students perceive the academic career path? What are women students' considerations in going for a postdoctoral training abroad? In what ways are women's considerations for and against pursuing an academic career shaped (constructed) by gender? To answer these questions, we choose a mixed-methods research methodology, combining a representative quantitative survey of all physics PhD students in Israel, with qualitative tools, conducting interviews with women PhD students and postdoctoral fellows (commonly termed 'postdocs').

In this chapter, we present the stories of these young women, while examining their considerations and trying to untangle the impact of the context – their personal, professional, economic, and family circles – on their decision-making. We start, however, by a review of the literature of this field.

Intersecting knowledge fields

The research that deals with the integration of women into academic careers is embedded in various academic disciplines, which are most commonly studied separately: psychology, sociology of education, sociology of organizations, gender studies, labour studies, economics, and more. The starting point for our study lies in the understanding that the answers to complex questions about barriers to women's integration within academia cannot be found in a single theoretical field or discipline. We must step beyond these traditional disciplinary borders towards interdisciplinary research, which can provide a diverse and multidimensional perspective to examine and answer the research questions.

The lack of interdisciplinary research among those dealing with career issues and the need to integrate perspectives have been highlighted as early as the 1970s (Van Maanen and Schein, 1979). Chudzikowski and Mayrhofer (2010) argue that the lack of interdisciplinary research in the field results in partial and biased understanding of careers, and a multilayered approach is needed.

In response to this challenge, we study career decision-making in physics as a multilayered and multidimensional phenomenon, combining in our analysis contextual, organizational, and individual variables and their intersection. In this context, looking at both structure and agency, we do not view careers as based solely on an individual's free choice, but rather on the intersection between individual choice, the roles and structure of physics as a field (in Bourdieusian terms, see later), gendered social expectations and norms, and family division of role and employment expectations. In the following subsections, we address the main theoretical background to our research and review the main concepts we use to analyse and explain our findings.

Gender and power

The understanding that power is an integral part of gender relations is central to feminist theory (Scott, 1986; Connell, 1987; Hartsock, 1987; Sheppard, 1989; Komter, 1991). Power is an integral part of Scott's definition of 'gender' (Scott, 1986). According to Scott, 'gender is a constitutive element of social relationships based on perceived differences between the sexes, and gender is a primary way of' signifying relationships of power. Changes in the organization of social relationships always correspond to changes in representations of power' (Scott, 1986: 1067).

The idea that social structure and processes are gendered was developed within second-wave feminist discourses dating from the early 1980s. Our theoretical starting point is based on the understanding that gender is a power structure in society, within family and in organizations (Kanter, 1977; Scott, 1986; Connell, 1987; Acker, 1990; Connell, 1990).

In her book *Gender and Power* (Connell, 1987), Connell characterizes the elusive way in which power works in social relationships. While particular transactions involving power are easy to observe, it is often difficult to see beyond individual acts of force or oppression to structure of power, a set of social relations with some scope and permanence. This elusive social power may be 'a balance of advantage or an inequality of resources in a workplace, a household, or a larger institution'. Connell uses the term 'gender regime' to analyse gender relations within the family (Connell, 1987, 1990).

Acker claims that gender is a structural feature of labour organizations (Acker, 1990). She proposes a theory of organizations and gender that follows Kanter (1977) and other scholars (Scott, 1986; Connell, 1987; Reskin and Roos, 1987) who claim that gender is a structural feature of organizations and not a characteristic feature of individuals bringing it to the workplace. Acker shows how deeply gender is embedded within organizations. She claims that 'the organizational structure is not gender-neutral, on the contrary, assumptions about gender underlie the documents and contracts used to construct organizations and to provide the commonsense ground for theorizing them' (Acker, 1990: 139). According to Acker, when we say that an organization is gender-biased, it means that advantage and disadvantage, exploitation and control, action and emotion, meaning and identity are patterned through and in terms of a foundational assumption of distinction between men and women (Acker, 1990: 146). The gendered structure of organizations operates to the advantage of men, as it keeps power and capital (social, economic, and cultural) in their hands.

Following these insights, we claim that gender acts as a power structure in academia, within the physics field as a space of knowledge production and as workplace, and within the family. As such, it produces and reproduces the career choices of young women.

Gender studies have changed in the last three decades in many ways, with growing recognition of feminist thought and activism in the Global South, and the development of queer studies and trans studies mainly in the US (Connell, 2019). Moreover, there is a wide understanding that gender should not be studied separately from other power structures. Therefore, the intersections of race, class, and gender have been widely accepted as an important aspect of feminist research and theory. The 'intersectionality' perspective has developed from the critique of the hegemonic 'White' second-wave feminist theory that was based on the experience of White women (Acker, 2006; Holvino, 2010). Scholars suggest new avenues to researching and publicizing the hidden stories at the intersections of race, ethnicity, gender, class, nation, and sexuality (Holvino, 2010).

Examining the intersection between gender and ethnicity, class, and religion is also very important in the Israeli context. The Israeli society is divided by two main axes – religion and ethnicity. There are social gaps between religious and secular Jews, between Jews and Arabs and between the geographical and social periphery and the centre, and those axes are interconnected and influence women in academy as well (Toren, 2009). The field of physics remains, however, a very homogeneous field, consisting mostly of men, White, Jewish, of medium–high social-economic background, coming from educated families (Eran Jona and Nir, 2019; Eran Jona and Perez, 2020).

We now turn to discuss aspects of gender and power within physics as a male-dominated field.

Physics as a social field

Physics is a masculine field, characterized by supremacy of a White male majority, with a masculine culture and a public image of masculinity dominating the field (Gonsalves et al, 2016). Women's integration into this field is relatively new, roughly since the 1980s, and the gender imbalance in the field is significant worldwide (Cochran et al, 2019). Following Bourdieu's theory (Bourdieu, 1977, 1986), we study physics as a social field. Social field is a patterned set of practices within a broader social space, which suggests competent actions in conformity with field-defined rules and roles. Within this perspective, it is often understood as a 'playground' or 'battlefield', in which actors, endowed with a certain field-relevant capital, try to advance their position.

As we discussed in Chapter 6, Traweek (1988) conducted an in-depth anthropological study of the world of physics, tracing the culture of high-energy physicists in one of the leading laboratories in the US. Traweek describes physics as a male domain and claims that masculinity is an organizing principle of the physics laboratory, as part of the common perception that science is a field of 'individual great men'. The field of physics is highly competitive. To be accepted as a member of the physics community, you have to perform that you are very committed, charismatic, highly motivated, dominant, and aggressive. Only the strongest and brightest manage to overcome these obstacles or, in her words, 'Only the blunt bright bastards make it' (1988: 88). This culture of physics as a male-dominated field is also prominent in later studies (Gonsalves et al, 2016; Lewis et al, 2016).

One of the examples of the gendered nature of organizations lies in the common gendered perception of the 'ideal worker', based on an image of a man who is free from the burden of taking care of children and family and can put his entire time into work (Acker, 1990). Acker claims that occupations and hierarchies assume a universal, intangible employee. This employee is, in fact, a man: a man's body, masculine sexuality, control over emotions, and minimal responsibility for reproduction. Images of the male body and masculinity that are dominant in organizational processes marginalize women and contribute to the preservation of gender segregation in organizations (see Chapter 6).

Acker's claims about the 'ideal worker' discussed earlier are certainly valid with regard to academic institutions, which were established and took shape in an era when most staff members were men and most of their spouses were housewives. Bagilhole and Goode examine Acker's claims in the context of academia. They argue that there is a standard model of the academic career (Bagilhole and Goode, 2001). This model is far from being gender-neutral. Instead, it is embedded in a masculine culture and based on a patriarchal support system. Bailyn (2003) studied the academic careers of senior women faculty in MIT. She argues that the belief, shared by women and men academics alike, is that the only route to being a first-rate scientist is to strive to be the ideal, perfect academic, for whom work is the total priority, and for whom there are no outside responsibilities and interests.

Benschop and Brouns, who examined gender aspects in academic institutions in the Netherlands, claim that the basis of the scientific quality standard lies in the 'Olympus' model of science, in which the dominant representation of the brilliant researcher is that of a young man at the top of the Olympus, away from the practices of daily life, rooted in the ivory tower of academia (Benschop and Brouns, 2003). This model is one-dimensional, gender-biased, not open to variance, and may alienate women and men who do not find themselves in it.

The perception of a profession as men or women is also influenced by the extent to which an occupation allows or does not allow combining family life with a career. This component was found to be a significant factor in the decision-making of the women in our study, who despite having completed their PhDs in Israel with an excellent grade, still pondered whether to pursue an academic career in science (Gofen, 2011).

Lamont and Molnár review the idea of 'boundaries' and how it explains various inequalities – class, race, and gender (Lamont and Molnár, 2002). With regard to the latter, their focus is on how gender and sexual categories shape expectations and work life. In this context, boundaries are defined as 'the complex structures – physical, social, ideological and psychological – which establish the differences and commonalities between women and men, shaping and constraining the behavior and attitudes of each gender group' (Gerson and Peiss, 1985). Violation of gender boundaries often leads to punishment and stigmatization in the workplace (Epstein, 2004). For a related discussion, see Chapter 6.

Sociological and organizational aspects of gender inequality

The attempt to understand the preservation of inequality in academia in general, and in sciences in particular – science, as well as technology, engineering, and mathematics (STEM), has given rise to a growing body of sociological and organizational research. Current explanations within these disciplines deal with conscious and unconscious processes taking place at the individual, organizational, and social levels.

These explanations can be divided into three categories: explanations concerning the preservation of the gendered power structure in society, including in organizations, that is also reflected in the field of science (for example, Connell, 1987; Acker, 1990); explanations concerning organizational structures, processes, and practices in academia that are biased towards men, their way of life, and merits (Atewologun and Tresh, 2018; Gvozdanovic and Maes, 2018; Eaton et al, 2020); and explanations concerning organizational culture and climate, which belong, to some extent, to the previous category, but are dealing more with conscious practices and behaviours that have a negative impact on women in academia (chilly climate, sexual harassment, and microaggressive behaviours towards women) (Johnson and Benya, 2018).

In the organizational context, studies have found that the inequality in academia starts with recruitment and screening processes (the 'similarity bias' effect, activation of criteria fitting men, such as the number of publications, and recommendation letters in which gender bias against women is present); continues further to working conditions and job characteristics (women receive lower wages than their male colleagues, less research resources, there is a gender bias in funding, in staffing resources, and so on); and further to unequal practices in the promotion process (promotion processes are biased in favour of men, and preservation of the 'scissors curve', in which the proportion of women decreases as the positions become more senior) (Long, 2001; Kulis and Collins, 2002; Wolfinger and Goulden, 2008; Gofen, 2011; Atewologun and Tresh, 2018; Gvozdanovic and Maes, 2018).

We now turn to discuss how gender inequality affects women's career decision-making.

Women's career decision-making

Psychological approaches to women pursuing non-traditional careers reveal that they face many obstacles and constraints that can limit or impede their career development. Those who wish to pursue an academic career in masculine fields must often overcome the absence of role models, weak self-efficacy beliefs, uncertain outcome expectations, along with cultural and institutional barriers.

Focusing on STEM fields in the academy, previous research reveals a high attrition rate for women before and during postdoctoral studies – a key period towards academic careers, where the numbers of women decrease dramatically (Goulden and Mason, 2009; Carmi et al, 2011; Gofen, 2011).

Choosing an academic path is a risky decision. The results from a meta-analysis of the research on 16 different types of risk-taking indicate that men participants are more likely to take risks than women participants, in almost all types of risks (Byrnes et al, 1999). The tendency for women to take fewer risks may be one of the explanations for the lower rates of women who choose the academic path, given its high demands and job insecurity.

A woman's decision to pursue a non-traditional career path, such as physics, could be also explained, following Social Cognitive Career Theory (SCCT) (Lent and Hackett, 1994; Kanny Sax and Riggers-Piehl, 2014), by their lower self-efficacy (on average). Low levels of self-efficacy beliefs among women pertaining to science have been implicated in contributing to the limited number of women earning postgraduate degrees and holding academic appointments (Lent and Bieschke, 1991; Chemers and Garcia, 2001; Grunert and Bodner, 2011; Falk et al, 2017). Other factors that may explain women's choice not to pursue career in physics include the imposter syndrome, mentoring and advising during graduate school, and the 'two-body problem' (Ivie and Chu, 2016).

These theories and explanations for decision-making are, however, partial and share the same problems as other social psychological theories of career decision-making – 'Trait Theory', 'The Developmental Model', and 'The Social Learning Theory' (see Hodkinson and Sparkes, 1997). All of them retain a strong sense in which decision-making is fundamentally an individual process, contains large elements of technical rationality, and remains within the influence or the control of individuals (Hodkinson and Sparkes, 1997).

In contrast to this, following Bourdieu and others, in this chapter we claim that women's career decision-making is affected by both their agency (the capacity for individuals to act independently of the social system and to make their own free choices) and social and context factors. While the psychological literature on decision-making may help to understand women's considerations in choosing an academic career (Nielsen, 2017), consent is never freely or neutrally given in situations of broader structural inequality (Beddoes and Pawley, 2014).

To better explain the power structures that influence women decision-making over an academic career, it is helpful to use Bourdieu's field theory (Bourdieu, 1984; Bourdieu and Passeron, 1990; Bourdieu, 1993); see also Hodkinson (1997). According to Bourdieu, agents do not continuously calculate according to explicit rational and economic criteria. Bourdieu uses the agency–structure bridging concept of field. A field can be described as any historical, non-homogeneous social–spatial arena in which people manoeuvre and struggle in pursuit of desirable resources. In his words, 'it is the state of the relation of force between players that defines the structure of the field' (Bourdieu and Wacquant, 1992). The 'players' within the field are various and they have different resources and power, which make up 'the relation of force'. For Bourdieu, each stakeholder brings capital to the field, which can be economic, social, cultural, or symbolic.

But, as we learned from decades of sociological studies, those various types of capital are all gendered, and gender operates as an asymmetric capital. Therefore, while masculinity operates in favour of men, femininity does not operate in favour of women (Weitz, 2001). Moreover, following Bourdieu's theory, the recognition of the limits of what is possible or what is not possible, encapsulated in the decisions of men and women, shapes their aspirations and career paths in a different way. Women's decision-making is shaped within a gendered power structure.

The Israeli context

In order to understand the research context, we would like to refer to Israeli-Jewish society, which has certain special characteristics. The proportion of Israeli-Arab students within physics graduate students is about one per cent. While this raises important questions, context relevant to this research is that of the Israeli-Jewish population.

Israeli-Jewish society is a familial society. The importance of family is manifested in the relatively high number of children per family, and in the great importance attached to the family in the life of both individual and community (Yair, 2020). The family is a central social institution in the individual life and part of the national strength. As a result of the centrality of the family, the woman in the Israeli-Jewish family is perceived, first and foremost, as a mother and a wife, and only then as a provider (Fogiel-Bijaoui, 1999).

Israelis marry on average at a relatively young age and have more children than in other Western societies. The average fertility rate per women in Israel is 2.7 children per woman (for all women excluding ultra-Orthodox Jews that hold a higher rate), a high rate compared with 1.7 average in Organization for Economic Cooperation and Development (OECD) countries (ICBS, 2020). While it is a conservative society, it is mostly open and acceptive to the LGBTQ+ community and to alternative family formations (same-sex families, single-parent families, and so on)

In Israel, most women are employed full-time, and most maintain a full-time employment history even during the years of raising children. The full-time employment rate (68.2 per cent) is much higher than the corresponding rate in other countries, such as Germany (54 per cent), Norway (60 per cent), the UK (57.7 per cent), and the Netherlands (41.6 per cent) (Guy and Shnider, 2019).

However, despite the large extent of Israeli women's participation in the labour market, and despite a discourse that promotes equality in the division of roles in the domestic sphere, the division of roles within Jewish-Israeli families is still unequal and in most cases the burden of raising children is not equal. As found in research conducted among Israeli fathers, even though there has been an increase in recent decades in fathers' involvement in household chores and childcare practices, the parental responsibility for the private sphere is still unequal. This sphere remains feminine. Men spend longer hours at work, and most of the household chores and childcare become their wives' burden (Anabi, 2019). These findings are in line with studies conducted in other Western countries; for example, in the US and Canada (see Doucet, 2015) and with the research on women working in male-centric domains, that experience daily battles as competing desires to both be a 'caretaker' at home and develop a professional career (Case and Richley, 2013). Family transformation in Israel is influenced by two contradictory trends – preservation versus innovation – each pulling in a different direction (Toren, 2009).

As for the integration of women in STEM fields in the Israeli academy, women are under-represented. The lowest representation of women is in physics: only 6 per cent of the academic staff in physics are females (Eran Jona and Nir, 2019); the rates are low also in mathematics and computer studies (11 per cent), physical and life sciences (13 per cent), and in engineering and architecture (14 per cent) (Mizrahi, 2015). According to Kark (2007), it is claimed that the under-representation of women in STEM fields in Israel is a consequence of the compulsory military service that is biased towards males, the social norms of familyism, and the gendered care responsibilities.

It was within this context that we conducted our research. Our research questions were the following: How do PhD students perceive the academic career path? What are the women student considerations in going for a postdoctoral training abroad? In what ways are women's considerations for and against pursuing an academic career shaped (constructed) by gender?

Methodology

As also described in Chapter 6, our research team is unique as it consists of a woman sociologist and a man physicist. The sociologist has had previous experience in studies of gender issues in the male-dominated military environment. The physicist, beyond his experience in theoretical particle physics research, gained some perspective of the cultural norms and the power structure of the field, by serving as dean of a physics faculty, a chairperson of an institutional promotion committee, and a member in various international advisory committees. Together we decided to step out of our respective disciplinary comfort zones and collaborate in a long-term research project on gender aspects in physics.

Mixed-methods research paradigm

In this study, we used the mixed-methods research paradigm – an intellectual and practical synthesis based on qualitative and quantitative research. The mixed-methods approach recognizes the importance of traditional quantitative and qualitative research but also offers a powerful third paradigm choice in order to provide the most informative, complete, balanced, and useful research results (Glesne and Peshkin, 1992; Denzin and Lincoln, 1998; Johnson and Turner, 2007). In parallel, to structure the research tools and analyse the research findings, we used feminist research approaches and theories that provide framework and tools for looking into women's lives (Reinharz and Davidman, 1992; DeVault, 1999; Krumer-Nevo et al, 2014).

Feminist approaches, which emphasize the need to voice the diversity of women's points of view, stem from the fact that their multiple positions were not included in the process of knowledge creation in the social sciences until the middle of the 20th century. Qualitative research methods provide a way to examine sensitive and private aspects of women's lives and constitute more appropriate tools to study power relations within the family and the workplace (Edwards, 1993).

The research discussed in both this chapter and Chapter 6 use mixed-methods methodology. Chapter 6, where we discuss the difficulties during PhD studies, was based mainly on quantitative data, supported by parts of the qualitative data. This chapter is based mainly on qualitative data collection and analysis, while concurrently recognizing that the addition of quantitative data and approaches into the research contributes to a rigorous understanding of the social phenomena that we study.

Qualitative data collection

In our series of surveys, which we conducted from 2018 to 2019 among all physics students in Israel studying for a BSc, MSc, and PhD degree, we found that these students belong to a hegemonic group in Israeli society, consisting mostly of men, White, Jewish, of middle-upper socio-economic background, and coming from educated families. Women are a minority among physics students in all degrees and their rate has hovered around 18 per cent in all three degrees in the last decade. As previously discussed, the number of Arab students, the main minority group in Israel, is very low: close to 1 per cent of the students or less. Ultra-Orthodox students, another underprivileged minority, are almost absent in physics. Another variable examined was gender self-definition: the vast majority of physics doctoral students defined themselves as either 'male' or 'female' (only three students defined themselves as 'other'), so we did not have a large enough group to explore this aspect (Eran Jona and Nir, 2019; Eran Jona and Perez, 2020).

Hence, although we fully acknowledge the importance of examining the intersections between gender and other social categories, as a consequence of the characteristics of physics in Israel as a field that lacks diversity in terms of religion, ethnicity, and status, we focused on examining the differences between women and men in physics as distinct categories.

To learn about the women's diverse experiences, we used interviews as the primary method of data generation. We conducted in-depth interviews with women physics PhD students studying at six research universities in Israel. Given the small size of the community (there are about 60 women physics PhD students in Israel in a given year), we had to reach out to many of them. We used personal networks to share a short introductory explanation about the study, after which participants who wanted to share their experiences got in touch with the research team. No student refused to be interviewed; some even agreed to be interviewed while on maternity leave. The interviews were conducted by Meytal, as the trained sociologist in the team, face to face in students' offices, labs, and sometimes in their homes or at a coffee shop, according to their request. Given the size of the physics community, we made sure that the interviewers had no personal or academic connection with any of the research participants. For that reason, Yosef, who is a physics professor and holds a position of authority within the physics community, did not take part in the interviews. Prior to conducting the interview, we provided a debrief explaining the research and the interview content. We used an open-ended protocol, including questions about academic track; overall experience during the PhD studies;

main difficulties and challenges; future plans and considerations to continue to postdoctoral training; pros and cons of an academic career; and family status and work–family challenges (for parents). Interviews lasted 60–120 minutes (most took about 90 minutes). All interviews were recorded, and the participants' names replaced with pseudonyms for confidentiality.

Overall, we interviewed 25 women physics PhD students. Mapping their research interests shows that 15 of the interviewees were experimental physicists and ten were theorists. The age of the interviewees ranged from 26 to 36. Most of them were married or in a relationship (21) and only a few were single (four). Twelve were mothers, with 1–4 children, mostly babies or toddlers. Although we were curious to learn about the diversity among the students in terms of sexual orientation and types of spousal relationships (heterosexual, same-sex, and so on), none of the interviewees defined themselves as belonging to the LGBTQ+ community. The model that repeated itself among all the interviewees is the heteronormative one. With regard to their spouse's occupation, ten were engaged in the fields of computers/engineering/exact sciences in the industrial or private sector, eight were PhD students in physics, similar to their spouses, and the others had various occupations.

To understand the next phase of the academic career, the postdoctoral path, we also conducted 13 interviews with Israeli women postdoctoral fellows. To pursue an academic career in Israel, postdoctoral research abroad is practically a necessary condition. Moreover, in order to obtain a tenure-track position in physics, two postdoctoral periods are often required; that is, a cumulative period of about four years abroad.

Once again, and similarly to the methods we used with the PhD students, we reached out to the postdoctoral fellows through our social network within the physics community and by using snowball methodology. Since they were interviewed during the postdoc period when the young scientists were abroad, conducting their research in universities and research institutes worldwide, most interviews were conducted online (using Skype). Interviews lasted 60–90 minutes. All interviews were recorded, and the participants' names replaced by pseudonyms for confidentiality. We also edited the data in a way that would preserve the anonymity of the participants. Most interviewees were in their mid-thirties, all of them were in a heteronormative relationship, and most of them (11) had children. The number of children ranged between one and four, with two being the most common. The majority were admitted to postdoctoral studies in leading institutions in the US and Europe. Since we followed their career in the years following the research, we can also note that the postdoctoral duration averages four years.

The aim of the interviews was to understand the individual and institutional factors impacting their career decisions, whether to pursue an academic career in physics or to leave the academy to a different career path. The interviews matched those for PhD students, and were based on an open-ended protocol, including questions about the academic track in physics; the personal experience during the postdoctoral training; main difficulties and challenges along the postdoc; the process of applying for a permanent position in the academy; family status and work–family challenges (for parents); and future career plans.

All interviews with PhD students and postdoctoral fellows were recorded, transcribed by the research team, and thematically analysed using ATLAS.ti software. Two researchers participated in the codification process and in the data analysis work, discussing the analysis for better interpretation of the data. The analysis of the data proceeded through three distinct phases held by both authors. Initially, a comprehensive thematic analysis was undertaken, encompassing all interviews, with the aim of discerning and scrutinizing recurring patterns of significance within our data set. These identified themes, among others, were subsequently incorporated into the coding framework as primary conceptual categories.

Then, second-order themes surfaced as we delved into the research questions about their academic career path and experiences, mainly focusing on the gender aspects of their experience and the decision-making process about their academic career. At this stage we used mainly qualitative research analysis based on the 'Grounded Theory' model (Denzin and Lincoln, 1998).

A final step in our analysis was the identification of the personal, academic, family, and other considerations in choosing a career in physics, with an emphasis on unique barriers and challenges for women.

Quantitative data collection

The second phase of the research included quantitative data collection. In order to have broader, representative data regarding all physics graduate students, male and female, we conducted a nationwide survey. Surveying both women and men allowed us to identify issues where there are no gender differences as well as those where there are.

The survey questionnaire was compiled by the research team in consultation with researchers at the American Institute of Physics (AIP), which has been researching student attitudes towards physics for a decade. The research questionnaire that was formulated is partly based on the tools developed at AIP for research in the field, while adapting it to the Israeli context and to the research questions that interested us (see, for example, Mulvey and Pold, 2020). The questionnaire included 106 questions, of which six were open-ended. The topics included: students' socio-demographic background, academic study track, attitudes regarding the academic environment, success indicators, combining family and studies, future employment expectations and intentions, desire to have an academic career, considerations in favour of and against postdoctoral studies, and aspects of discrimination and sexual harassment during the academic studies. Some findings of the survey are beyond the scope of this chapter and are reported elsewhere (Eran Jona and Nir, 2020).

Physics graduate studies are only possible at a few universities in Israel. Therefore, to conduct the research, we reached out to the Israeli Physical Society (IPS) for partnership and support in our study. Through the IPS, we were able to reach the six physics faculties in the Israeli universities that have a PhD track in physics: Bar-Ilan University, Ben-Gurion University, Hebrew University, the Technion, Tel Aviv University, and Weizmann Institute of Science. Together with the IPS, we approached the six physics deans and asked for their help in distributing the survey. Indeed, all university deans forwarded the survey request to their PhD physics students. The deans also shared with the research team their data about the number of active students by gender. Following the data collection, we were able to have the final numbers of physics PhD students in Israel by gender: n=404 students, of whom n=64 women and n=340 men (in 2019).

To enhance the response rate, we promised all students full anonymity, distributed the survey through the faculties' mailing lists, and reached out to students to encourage them to answer the questionnaire. We also gave all participants a $15 card to buy books as a thank-you gift. We managed to receive a very high response rate: 66 per cent of the overall population of students in the country (267/404), with an even higher response rate – 94 per cent – for women (60/64), and 61 per cent response rate for men (207/340). We received answers from students in all six universities. The population size and response rate by institution are presented in Table 4.1. The maximum error for the entire population is 3.6 per cent: among women 3.2 per cent and among men 4.3 per cent. Due to the representation of

Table 4.1: Physics PhD students in Israel: the overall population and the number of respondents by institution

Institution	Students	Respondents	Response rate
Weizmann Institute	109	71	65%
Bar Ilan University	67	50	74%
Hebrew University	65	40	62%
The Technion	65	39	60%
Tel Aviv University	64	39	60%
Ben Gurion University	34	28	82%
Total	404	267	66%

women in the sample, the total data of the students were weighted by gender. Data analysis was performed using variance to proportions analyses, given small populations.

The main basis for the findings presented in the next section was qualitative content analysis of the interviews, which gave rise to a coherent picture of the main considerations taken by women physicist PhD students when they decide whether to continue for a postdoc. Our insights were further supported by the quantitative analysis of the survey data.

Findings

The postdoc as a 'deal'

Exploring the students' expectations of an academic career illustrates their image of the field. Within our nationwide survey, we asked all PhD students that state they would pursue an academic career (143/248) the following open question: 'If you are interested in an academic career, please elaborate why?'

Their answers were somewhat surprising. We found that love for physics and a deep interest in this field were the most common answers (66) (inter alia, loves research, loves exploring, loves basic research, this is my dream, it fits my character). The next common answer (33) was the academic freedom (the ability to conduct my own research, independence, no bosses or customers). The third common answer (21) was related to the work conditions (favourable conditions, tenure, job security, prestige, social status, and leadership capacity).

All three reasons were also raised in the interviews. We thus conclude that for many PhD physics students the main benefits of the academic career 'deal' are the ability to engage in scientific research in a fascinating field, intellectual freedom to explore and be creative, independence in choosing what and how to do research, freedom from bosses or clients, and working conditions that guarantee employment stability and (reasonable) economic well-being. The deal does not include quick enrichment, but it includes the prospect of groundbreaking scientific discoveries (and worldwide fame alongside them), as well as prestige that comes with being part of the exclusive club of the intellectual elite.

Considering these career benefits, we found that, at the crossroads of pursuing a postdoc, the academic career is considered as a 'deal', which has three main components: personal-marital, professional-occupational, and financial. Many of the young women in our research are realistically examining the components of this 'deal': what it offers them and what cost they will have to pay. In accordance with these considerations, the decision is made. This does not imply that the considerations are all 'rational' or 'objective'. Undoubtedly, the decision involves feelings and thoughts, realistic and unrealistic expectations, perceptions of academic institutions and labour market, and aspiration of professional and personal future, but the bottom line is that all of the above are merged into one informed decision, whether to go for a postdoc abroad as a necessary step for an academic position, or to quit the academic race at the current stage.

Embarking on a postdoctoral career is a significant, even dramatic, step in the lives of young women (and men), and requires a will to make significant changes in many aspects of life for a long period of time. The rules of the social field demand that, in order to obtain a tenure-track position in physics, two postdoctoral periods are often required; that is, a cumulative period of about four years abroad. Therefore, it has personal, family, professional, and financial implications for both women and men, and occurs under conditions of uncertainty and job insecurity.

On the personal and family level, relocation to a foreign country is required, which includes in many cases the need to integrate children into new schools and kindergartens while learning a new language, and integrating into a new social and cultural environment. This could be particularly difficult for single women, with no spouse or family to rely on.

On the professional level, the candidate must find an academic mentor and an institution willing to host her for the postdoctoral period. This task requires talent, self-marketing skills, willingness to travel abroad, and an effort to become acquainted with suitable scientists and institutions. During the postdoctoral period, the candidate is required to prove herself again, publish, make a good impression on the relevant professional community, and prove her capability as an independent scientist. After all these efforts, the postdoctoral researcher is still not guaranteed to get an academic job, as competition for jobs is high, availability is low, and the chances are unknown.

On the financial level, there is the concern for making a living. The scholarship during the postdoctoral period is significantly lower than physicists' average wage in the labour market. It does not include social benefits and accrual of future rights, such as pension and education fund. Most scholarships are designed to allow a single person to live a modest life, and usually do not suffice for a family. The costs of living abroad for an Israeli family with young children may be significantly higher than the average scholarship. It means they have to fund postdoctoral studies via savings or family support, in a time of their lives when they are expected to be financially independent. Moreover, relocation abroad may impair the spouse's income, employment continuum, and skills. For some professions, it is hard to find a parallel job abroad (for example, lawyers or military officers); in some cases, immigration-related restrictions do not allow the spouse to work.

In order to learn about the characteristics of the academic track in physics and its requirements, the students must gather information from colleagues, friends, and supervisors. Because of the minority of women in the field, all universities in Israel take various actions to encourage women to go on a postdoctoral fellowship, by providing among other things detailed information about the track. Therefore, women we interviewed were aware of these components when making the decision whether to go on postdoctoral studies abroad.

When D, a young single woman in the last year of her PhD studies in physics, was asked if she plans to go on a postdoc abroad, this was her answer:

> 'The post seems to me as a kind of interesting experience, but I think you need to think about how you see life afterwards, because the post is something that has a lot of costs, you have to move to another country, it's not completely trivial with a family, and in terms of life afterwards there is a lot of uncertainty, and there is an expectation of a few more years of job insecurity, the chance of getting a job is very low, and there are also things that I think I will suffer a lot from, for example writing research grant proposals and the need for self-marketing.'

D's answer includes all the elements of the 'deal': personal, professional, and financial. She had already considered it and her decision was not to go on a postdoc. The considerations that mostly influenced her decision, as she said later in the interview, were personal and financial. Economically, she comes from a low status and will find it difficult to wait another four years at a low salary and without job security, until the long-awaited job. Personally, she is single without a family and is afraid to leave the country alone, without her family support. She was also concerned that she might not have some of the skills required to succeed in academia. So, weighing up all the elements of this path she decided that an academic career is not the right 'deal' for her.

The gendered aspects of the deal

While both women and men examine the 'deal' terms at a similar stage of their lives, it is clear that among women, the gendered power structure creates different expectations and extra hurdles, which make their decision to go for a postdoc more challenging.

Gendered power structure influences women's academic careers in physics in numerous ways. First, women face unequal competition in physics as a masculine field. Second, couples prioritize the man's career over the woman's career. Third, a postdoctoral career path is socially perceived as a disruption of the gender order. Women justify this non-normative path by demanding of themselves exceptionally high standards of academic excellence. We claim that these standards of excellence operate as a hidden component within the gender regime that justifies women's decision to go for a postdoc.

Unequal competition in physics as a masculine field

As discussed in the literature review, physics is a masculine field, characterized by supremacy of a White male majority, with masculine culture and masculine public image of the field. Thirty years after Traweek's work (Traweek, 1988), the field of physics is still masculine and highly competitive in many Western countries, including Israel. The gendered labour market in the physics field is clearly reflected in both our survey findings and the interviews. Over the last decade, women constitute only 18 per cent of the BSc, MSc, and PhD physics students in Israel and there are no signs of positive change. At the faculty level nationwide, women constitute only 6 per cent of the overall staff in all physics faculties (Eran Jona and Nir, 2019).

The marginal position of women in the field of physics is evident not only quantitatively, but also qualitatively, in the women's experiences (see Table 4.2). Most women PhD students reported having experienced gender-based discrimination during their studies, compared with a small minority of men. Moreover, 1 in every 5 women reported being sexually harassed during their studies, compared with 1 in 40 among men. Of these, half were harassed more than once. Only a minority of victims reported it to official entities. Similar findings in this regard were reported in Barthelemy et al (2016), based on interviews with 21 women in graduate physics and astronomy programmes. For a related study, see Aycock et al (2019).

As we found in the interviews, in physics as a male-centric culture, the notion that women are 'others' plays against them.

In her interview, a doctoral student who came to study in Israel from Germany (married without children), described her experiences, and that of an Italian woman postdoc (single) in her research group, as an ongoing challenge:

'We work together in the group with nine men and some people get confused between us, and it's strange – as if they made us one, one woman and didn't see our individuality. It bothered us a lot. Or in big

Table 4.2: Survey results on the issues of gender-based discrimination and sexual harassment

Topic	Women Yes/Total (%)	Men Yes/Total (%)	P
Have you experienced gender-based discrimination?	34/54 (63%)	5/167 (3%)	<0.01
Have you been sexually harassed during your studies?	13/59 (22%)	5/204 (2.5%)	<0.01

conversations, sometimes colleagues wouldn't let us finish talking. We have to speak more forcefully, to speak more loudly or to be more vocal, we thought together how to speak so that we would have more voice. There were times when the imbalance in the team worked against us.'

The gender disadvantage becomes more prominent when they become parents. While women are expected to become the main caregiver, their men colleagues are expected to follow their career as they had done before they became parents. Timing works to the detriment of women since, while they need flexibility in order to raise young children, they have to prove themselves to a highly demanding system that does not stop for a moment (Ceci, 2010), a system in which there is no such thing as a 'good' time to have children.

Although many of the women we interviewed said that they strive to implement an egalitarian model of childcare at home, most of them also report they spent longer hours in childcare and child-related work compared with their partners, and that this comes at the expense of their studies.

These findings are supported by the survey data indicating that, despite a prominent presence of an egalitarian ideology among physics student families, expressed by women's and men's desire for an equal distribution of roles, it is evident that women carry a greater burden of family work. First, they take a longer maternity leave, a period of time that impedes their studies: most women take maternity leave of four months or more, compared with very short parental leave used by the male students (see Table 4.3 for the full survey results). As apparent from the interviews, women are also going through the pre-birth period, during which many women need to undergo various examinations, and sometimes require medical care and observation, which take time and require a lot of attention (a reality that may repeat itself after maternity leave as well).

Second, women, more than their male colleagues, reported that due to parenthood, they adopted a more flexible work schedule, and that they

Table 4.3: Survey results on the following question: what was the duration of the parental leave you took after the birth of your last child?

Duration	Women n (%) (n=21)	Men n (%) (n=93)
≤ week	1 (5%)	61 (66%)
week–month	0 (0%)	28 (30%)
1–3 months	5 (24%)	4 (4%)
≥ 4 months	15 (71%)	0 (0%)

(x^2=92.67, $p<0.00001$)

learned to be more productive in their studies (see Table 4.4 for the full survey results).

Third, examination of the role distribution structure in these families indicates that, although the fathers are engaged in childcare, there is still gender inequality at home (see Table 4.5). 100 per cent of women reported that they are responsible for taking care of their children's needs, either as the primary carer, or sharing childcare with their spouse equally. In contrast, while the majority of men reported that they share the childcare responsibility with the spouses, more than a quarter reported that most childcare responsibilities are imposed on their spouses, and only a small minority reported to bear most of this responsibility.

Moreover, although most women and men reported that they share household chores, no women are free from this burden (the remaining 33 per cent reported that the household burden mostly lies on their shoulders), while men either share the burden witʰ their spouses or it is mainly imposed on their spouses (see Table 4.5 for the detailed results).

These findings are not surprising. They are manifest in studies about time spent on housework in Western democracies for decades. According

Table 4.4: Survey results on the following question: how did your course of study change as a parent?

	Women n (%) (n=18)	Men n (%) (n=89)	P
Reduced time for study	15 (83%)	64 (73%)	
More flexible schedule	12 (67%)	42 (48%)	<0.05
More efficiency	8 (44%)	24 (27%)	
Reduced pace of studies	7 (39%)	30 (34%)	
No change	1 (6%)	12 (14%)	

Table 4.5: Survey results on the issues of childcare and household

Topic	Women n (%)			Men n (%)			P
	Student	Equally	Spouse	Student	Equally	Spouse	
On whom lies the main responsibility of childcare?	12 (57%)	9 (43%)	0 (0%)	5 (5%)	61 (66%)	26 (28%)	<0.01
On whom lies the main responsibility of household?	7 (33%)	14 (67%)	0 (0%)	11 (12%)	61 (64%)	23 (24%)	<0.01

to the American time use survey, women spent an average of 2 hours and 15 minutes a day on housework, while men spent 1 hour and 25 minutes (Bureau of Labor Statistics, 2016). Women continue to take the primary responsibility for home and family even in some of the most gender-equal countries (Seierstad and Kirton, 2015).

This gendered role division at home is reflected in the academy, by a more significant presence of men at physics labs and offices, in terms of time allocation. Most male physicists (even if they truly believe in gender equality) do not practise parenthood in the same way as women. Though most PhD male students declare that parenthood affected their studies, parenthood is not as significant a variable in their lives as employee, as they follow the social expectation that the family will be pushed aside due to their career demands. At the same time, the usual expectation from the Israeli women is to do both, to be both dedicated mothers and career women.

The interviews show that women's greater commitment to family makes it harder for them to succeed in their studies within a male-dominated culture. The young women report being discriminated based on the normative assumption that mothers are less competent and committed than other types of workers, as was documented in previous research addressing the 'motherhood penalty' phenomenon (Benard and Correll, 2010).

T, a mother of a 2-year-old girl who was pregnant at the time of the interview, told us about the difficulty of combining studies with motherhood. T aspires to an academic career but, at the same time, she gives high priority to her family. She tries to live up to the social expectation of her 'doing both'. She finds out that this situation puts her at a disadvantage in her daily competition with her male counterparts. Her commitment to family, or what she calls 'this problem', is her problem, and not a problem for her colleagues, all of them male and free from having to live up to the expectation of being the main caregivers for their children. T describes it in a tone of acceptance, but also criticizes it:

'My family, my husband, my marriage, and my children are very important for me, they are very high in my priorities … I feel the gender differences (compared with male colleagues), mainly since I have much less time to work than my friends from the lab, and it becomes a big gap … it's like you are competing against those to whom you compare yourself, all the time … It's hard to combine motherhood with anything that is career related, not only in academia, but the competition is a competition with men who don't have this problem.'

As a young mother, the difficulty of combining career and motherhood requirements become clear to T. Taking care of her 2-year-old daughter takes precious time and harms her ability to successfully compete with other students, her lab colleagues, whose time is at their disposal and who are less challenged by family demands.

Her experience reflects many young mothers we interviewed. Their experience is structured within a gendered labour market, where men and women must live up to the same expectations at work, in the public sphere, but to different expectations in the private sphere. After becoming mothers, many of them understand that the competition, in which they are and will be required to compete, is unequal. This understanding leads some of them to quit the academic career race.

Prioritizing family and the husband's career

In Israel, in comparison with other Western democracies, students start academic studies relatively late and marry at a relatively young age. Therefore, PhD students' family situations in Israel have unique characteristics.

Based on the PhD student survey, men and women PhD students in physics are in their early thirties. The average age is 29.7 among women and 31.8 among men (probably because of the longer compulsory military service for men in Israel). Most women and men are already married or in a relationship, and more than a third of them already have children. Moreover, about 23 per cent of men and 17 per cent of women PhD students have two children or more (see Table 4.6 for details).

This picture is clearly reflected in the interviews. Most of the PhD students we interviewed (21 out of 25) already had spouses, half of them (12) were already mothers of young children at the time of the interview, while even the single women declared their desire to become mothers in the coming years.

Most spouses of the women students we interviewed were working in STEM fields (for example, engineers and computer scientists) or were graduate students in STEM. Most of the women students described their spouses as mostly supportive in their career aspirations and decisions, wanting and willing to help them succeed. In the survey, many students described their spouse as one of the factors for their own success (89 per cent of women [42/47] and 80 per cent of men [124/155], $p<0.01$). However, the interviews clearly indicate that, once there is a spouse, career and family

Table 4.6: Family status of the survey respondents

	Women n (%) (n=58)	Men n (%) (n=189)
Married	37 (64%)	132 (70%)
With children	20 (34%)	77 (41%)
≥ 2 children	10 (17%)	44 (23%)

considerations become intertwined and therefore more complex. Career becomes 'spousal' in the sense that a decision made with regard to the man's career affects the woman's career, and vice versa. The couple is considering the impact of their choices on the entire family.

The interviews with both PhD and postdoc women students prominently showed that many women give significant weight to the implications of going for a postdoc on their spouse's career. Women are preoccupied with the questions: Will my spouse be able to find a job or a postdoc abroad (in view of language- and visa-related barriers)? Will he be able to find work abroad that fits his skills and ambitions (depending on his professional characteristics, job availability, his ability to adapt to a different job market, and so on)?

A clear example of a 'spousal' career and interrelated career considerations emerges from the story of S, married with two toddlers. S and her husband are both PhD students in the last stage of their studies. S found it difficult to separate her career aspirations from her partner's, which she presented as interdependent. The interviewer tried to refine the differences and understand what she wants:

Q: "I'm trying to understand how you see your career, if you didn't have any limitations, where would you like to see yourself?"

A: "Now, it seems that my doctorate is going to be more or less successful and I'll have good results, so yes, I would like to continue in academia and do a postdoc, and I know that there are some institutions in Europe … they are looking for people (in our field) and we can both get work there, some sort of a postdoc, so that could be nice. But now it comes to my husband and if he finds a place that he will really love, and there won't be a place for two people there, I'll go look for a job in high-tech or something else, and that would also be perfectly fine, and if I won't find a job in high-tech, I will be a teacher and it will also be fun … It won't be as interesting as research, but it's a job."

S subordinates her desires to go for a postdoc in physics to those of her spouse; his career is being given a clear priority within the spousal relationship, although her PhD was good and she recently won an excellence award.

In another case, the counterweight to postdoc was the spouse's desire to stay in Israel. BG is freshly married, and pregnant with her first child. She told us that her spouse supports her, but at the same time he does not want to leave Israel and is very connected to his country and his family. BG eventually decided to leave academia. It is impossible to determine if it was her spouse who affected the decision against postdoc, but it was clear that his will had a significant weight on her decision.

Reinforcement of the findings that emerged from the interviews is found in the survey. While both men and women reported that their spouse's employment considerations play a key role in their decision whether to go for a postdoc, women have given more weight to this issue. Similarly, a higher percentage of the women indicated their spouse's ability to find a job abroad as a key consideration, and a higher percentage of women noted the difficulties involved in relocating abroad with the spouse and family as a key consideration (see Table 4.7 for the full survey results).

Thus, it seems that even if the male spouse is supportive and willing to follow the woman abroad, women give great weight to their spouse's career, desires, and preferences. This reflects, inter alia, the gendered power structure that exists within society as well as in the job market.

The priority and precedence given to the man's career in many families could be based on the understanding of their better chances for higher wage and promotion (on average) in the job market. It was surprising to find that, at this early stage, when most students live on a modest subsistence scholarship, there are already financial gaps in income, in favour of men. The majority of women PhD students indicated that their spouse's income from study or work is higher than their income, compared with a minority of male PhD students (see Table 4.8 for the detailed results).

Table 4.7: Survey results on the following question: if you are considering going abroad for a postdoc, how central is this consideration in your decision?

	Women n (%) (n=41)	Men n (%) (n=127)	P
Spouse's employment possibilities	39 (95%)	114 (89%)	
Spouse and family relocation	38 (93%)	107 (84%)	<0.05
Putting the spouse's career on hold	37 (90%)	99 (78%)	<0.05
Funding	37 (90%)	126 (99%)	<0.05

Table 4.8: Survey results on the following question: compare your spouse's income to yours

	Women n (%) (n=37)	Men n (%) (n=134)
Higher than mine	25 (67%)	37 (28%)
Similar to mine	8 (22%)	27 (20%)
Lower than mine	4 (11%)	61 (46%)
Has no income	0 (0%)	9 (6%)

(x^2=24.48, $p<0.00001$)

Postdoctoral career path as a disruption of the gender order

As explained previously, in the Israeli context, there is a requirement to undergo postdoctoral studies abroad; that is, the student must leave the country for a prolonged period of professional research and development. (The probability of getting an academic position after a postdoc in Israel is much lower.) Although the common view in contemporary educated circles is that the job market is open and equal for women, the interviews reveal a much more conservative view. It is evident that in many cases the social and family environment perceives postdoctoral studies as an ambitious and non-normative path for women. As was described by a few interviewees it is common for women to follow their spouses for a period of work or studies abroad, but the opposite model is still considered non-normative and is perceived as 'feminist' and challenging the common social order, in which the male career is the lead.

TS is about to embark on a prestigious postdoc in the US with her husband. She describes the postdoctoral path as non-normative for a woman, which is why it has to be negotiated with her spouse and justified against the family system:

'I think a postdoc abroad takes a heavy toll. Usually, the husband is older and has a job, and does not want to leave. The easy cases are when the husband also goes for a postdoc, or can work abroad and, in such a case, he wants to leave. Even if the husband is supportive, the broader family wrinkle up their noses and put pressure on me (not to go on a postdoc abroad). If it were possible to do a postdoc in Israel, it would be much easier for women.'

M, single, PhD student, believes that women are less likely to go for a postdoc because it is a deviation from the conventional structure of gendered power relations. In the accepted social order, the man's career is the significant one, and not the other way around:

'I think that the cultural perception, at least in Israel, is that the woman will follow the man; that is, if the man has to relocate due to work or studies, it is perceived as more natural for the woman to follow him.'

M says the postdoc issue has come up in her previous relationship, and although her former boyfriend's attitude regarding this issue was positive, his social environment was against this move and regarded it as non-normative. The idea of him, a man, relocating abroad for the benefit of his girlfriend's career, was met with criticism and astonishment by his colleagues.

In a few of the interviews it was clear that the priority given within the family to the male spouse's career restricts women from embarking on a postdoc and limits their choice of academic career.

Moreover, three Israeli single PhD students whom we interviewed claim they are refraining from embarking on postdoctoral studies due to the concern that they may impair their chances of getting married in the future or create potential limitations for a future (spouse's) career. This gendered power structure is so profound that it even affects single women, who do not have a spouse and children.

Self-expectations for excellence: a hidden component in the gender power structure

In a gendered workforce, when women compete over an academic career in a masculine field, what justifies their decision to go on a postdoc abroad, or to embrace the academic 'deal'?

Our findings indicate that, to justify this deviation from the 'gender order', women are pushing themselves to excel. A few women stated that it is considered obvious when a man goes for a postdoc that his wife goes with him, even if it requires that she gives up her career. In contrast, for a woman to embark on a postdoc with her partner joining her, special conditions must be met. One of the unspoken conditions, mentioned repeatedly during the interviews, is being an 'excellent' student: a concept which is usually understood in traditional 'masculinist' terms. It was particularly interesting to learn from those who decided to leave the academic path. This is how BG, married and pregnant, explains why she decided to leave academia for a job in the industry after completing her PhD:

> 'Women are also affected by their partner, not that men are not, but to a certain extent, when a man thinks about going for a postdoc, his wife is excited to follow him … it's an adventure, (on the contrary) a woman waits to hear the man's opinion, and if he says no, then there should be a really good reason, for example, when (your) doctorate is brilliant and the supervisor wants you to go for postdoc really pushes you … then, maybe then.'

As BG explains, to deviate from the norm, the woman must be 'excellent' – a term usually understood as associated with masculine characteristics and judged in that vein. It is not enough for her to be a good or even a very good student, and she probably cannot afford a postdoc if she is an average student. The 'excellence' must be reflected in numerous aspects that are interconnected: her academic achievements; her supervisor's evaluation of her; others pointed also to the professional group perception of her academic potential; and her ability to receive a postdoc offer from a top institute.

Findings indicate that the 'excellence' has two functions, in the public sphere and in the private sphere. In the public sphere, the ascription of 'excellence' allows the students to feel worthy to face professional competition against their male colleagues. In the private sphere, it justifies (to the spouses and to the student herself) their choice of a non-normative career path for a woman.

A very clear example of 'excellence' as a justification for disruption of the gender order (in both the public sphere and the private sphere) comes out from the story of A. A, recently married with a baby, said she had already decided when she started her PhD studies that she would not pursue a postdoc in a place that is not scientifically excellent. She says that, in the beginning of the relationship, her husband and she agreed that she would pursue postdoctoral studies if she were to be accepted into a prestigious institution. In return, he promised that he would be willing to leave his job to support her career. This is her answer to the question of whether she would like to embark on postdoctoral studies abroad:

'Yes, very much! But it is contingent upon me getting a good postdoc! I mean, not a postdoc from the University of Nowhere, I don't know, something like that, it should be a good postdoc! Because basically, my husband will come with me and will have to take a leave without pay, which is also not so trivial at work … I won't drag the entire family if it's a postdoc that will get me nowhere, you know, it should be a good and lucrative postdoc, so I would have some motivation to return to Israel. I really want to be in academia, and I think that science is just the best thing there is, in my opinion, and it's something I want to do all my life.'

Another example arises from the story of V, married and mother of three, who is about to graduate. She shares her doubts about the future. She claims her spouse supports her postdoc aspirations; however, if they relocate, he would have to give up a job that is a significant part of his life. Therefore, for V, going for a postdoc takes a high price for her partner's career. After some deliberation, V decided to do her first postdoc in Israel. Only if it is successful, she will go abroad for another period.

As found in previous studies, women must demonstrate stronger abilities than men in order to be recognized as equally good. In male-dominated disciplines, for women to be considered good, worthy of employment and promotion, they must often be better than their male peers (Heilman and Haynes, 2008; Kaatz and Carnes, 2014). In this research we find that this structure exists not only in social perceptions of women, but also in their self-expectations, affecting their decision-making regarding an academic career in physics.

Summary and discussion

In this study, we used a nationwide survey, as well as in-depth interviews with women physicists, as a robust empirical base to explore women decision-making in physics at the crossroad of academic career. Within physics as a social field, we study career decision-making as a multilayered and multidimensional process. The concept of physics as a social field follows Bourdieu's work (Bourdieu, 1986) that is particularly concerned with how social inequality is perpetrated and maintained through the use of capital (economic, cultural, social, and symbolic).

The theoretical novelty that we suggest is viewing this process as a 'deal', which involves contextual, organizational, and individual variables, and their intersection. Young women are examining the components of this deal: what it offers them and what prices they will have to pay, but their decision is made within a gendered power structure. Studying both context factors and agency, we find that the academic career in physics offers a 'deal', which has three main components: personal-marital, professional-occupational, and financial. Young women are realistically examining the terms of this 'deal', what it offers them, and what prices they will have to pay. The decision is made in accordance with these considerations but, contrary to men, women are operating within a gender power structure that navigates their decisions in a different way.

While both women and men consider the 'deal' terms in a similar stage of their lives, among the women, the gendered power structure creates different expectations and extra hurdles, which make the decision to pursue an academic career and to go for a postdoc more challenging. Our findings reveal the multiple and hidden ways in which gender operates as a power structure in the labour market within physics as an academic field, in the family within the private sphere, and in the social norms and expectations within society, putting up a barrier to women's academic careers.

This latent power structure influences women's decision-making and experiences in several ways. In the academic field, it produces unequal competition in a male-dominated playing field, where women struggle to succeed as physicists and as mothers, but are viewed as less-devoted workers because of their parental commitment. Within the private sphere, women carry a greater share of the childcare and family work, and, moreover, give priority and precedence to their husband's career and preferences. In the social sphere, choosing a demanding academic career is seen as a non-normative trajectory for women and as disrupting the gender order.

Women justify this non-normative path by raising their self-expectations for excellence. They feel that they must excel in their research and exhibit exceptional achievements. We claim that excellence operates as a hidden mechanism within the gender regime, that can justify women's decision to go for postdoc, but can also operate as an exclusionary mechanism that prevents many talented young women from choosing an academic career in physics. We should stop thinking about women as giving up the academic career ('the leaky pipeline' discussion – see, for example, Blickenstaff, 2005; Cimpian et al, 2020), but rather as choosing career paths that align better with the gender regime within diverse social and cultural contexts.

Our acquaintance with the physics culture in the US and in Europe, as well as our close collaboration with the community that conducts research on gender in physics in the European academy (particularly the GENERA[1] network), lead us to believe that many of the main findings of our research and our theoretical arguments apply not just to physics in Israel, but also more universally, at least in the context of physics in Western countries. The unique features of Israeli society are helpful in making some aspects of examining the physics career as a deal stand out more clearly. Further research is needed, however, to support these findings and to explore hidden barriers to integrating women into academic careers, in physics and other scientific disciplines where they still constitute a token minority.

If we want the academy to be more gender-balanced, so that women are no longer a token minority, we should tackle the many obstacles women face, both in physics as a field and within the family circle. While academia does not have the ability to significantly influence the division of roles within the family, it does have the ability to influence the design of the scientific career path. To bring more talented women to physics we should make physics as a culture and a career structure more appealing for women. We may do so by challenging the 'normative' career path that requires extensive international postdoc experience, travelling to conferences and schools and long working hours, as a condition to succeed, as it harms women unequally. If academia does not act in this direction, it will lose talented women to the global tech companies, which in recent years have made intensive efforts to change the gender balance among their employees and integrate more women into diverse work teams.

Acknowledgements

We thank Sharon Diamant-Pick for her help in conducting the survey and in analysing the interviews.

Yosef Nir is the Amos de-Shalit chair of theoretical physics. This research is supported by grants from the Israeli Ministry of Science and Technology, from the Estate of Rene Lustig and from the Estate of Jacquelin Eckhous.

Note

[1] Gender Equality Network in Physics in the European Research Area (www.genera-netw ork.eu/).

References

Acker, J. (1990) 'Hierarchies, jobs, bodies: a theory of gendered organizations', *Gender and Society*, (2): 139–58.

Acker, J. (2006) 'Inequality regimes: gender, class, and race in organizations', *Gender and Society*, 20(4): 441–64.

Anabi, O. (2019) 'Fathers' involvement in the domestic arena and ideologies of masculinity in Israeli society', Doctoral dissertation (in Hebrew), Israel: University of Bar-Ilan.

Atewologun, D., Cornish, T., and Tresh F. (2018) 'Unconscious bias training: an assessment of the evidence for effectiveness', retrieved from Equality and Human Rights Commission, Research Report 113, Available from: www.equalityhumanrights.com

Aycock, L.M., Hazari, Z., Brewe, E., Clancy, K.B., Hodapp, T., and Goertzen, R.M. (2019) 'Sexual harassment reported by undergraduate female physicists', *Physical Review Physics Education Research*, 15(1).

Bagilhole, B. and Goode, J. (2001) 'The contradiction of the myth of individual merit, and the reality of a patriarchal support system in academic careers: a feminist investigation', *European Journal of Women's Studies*, 8(2): 161–80.

Bailyn, L. (2003) 'Academic careers and gender equity: lessons learned from MIT', *Gender, Work and Organization*, (10): 137–53. DOI: 10.1111/ 1468-0432.00008

Barthelemy, R.S., McCormick, M., and Henderson, C. (2016) 'Gender discrimination in physics and astronomy: graduate student experiences of sexism and gender microaggressions', *Physical Review Special Topics: Education Research*, 12(2), Article 020119. https://doi.org/10.1103/PhysRevPhys EducRes.12.020119

Beddoes, K. and Pawley, A.I. (2014) 'Different people have different priorities: work–family balance, gender, and the discourse of choice', *Studies in Higher Education*, (39)9: 1573–85. DOI: 10.1080/03075079.2013.801432

Benard, S. and Correll, S.J. (2010) 'Normative discrimination and the motherhood penalty', *Gender and Society*, 24(5): 616–46.

Benschop, Y. and Brouns, M. (2003) 'Crumbling ivory towers: academic organizing and its gender effects', *Gender, Work and Organization*, 10(2): 194–212.

Blickenstaff, J.C. (2005) 'Women and science careers: leaky pipeline or gender filter?', *Gender and Education*, 17(4): 369–86.

Blue, J., Traxler, A.L., and Cid, X. (2018) 'Gender matters', *Physics Today*, 71)3): 40–6.

Bourdieu, P. (1977) *Outline of a Theory of Practice*, Cambridge, MA: Cambridge University Press.

Bourdieu, P. (1984) *Distinction: A Social Critique of the Judgement of Taste*, Cambridge, MA: Harvard University Press.

Bourdieu, P. (1986) 'The forms of capital', in J.G, Richardson (ed) *Handbook of Theory and Research for the Sociology of Education*, New York: Greenwood, pp 241–58.

Bourdieu, P. (1993) *Sociology in Question*, London: Sage.

Bourdieu, P. and Passeron, J.C. (1990) *Reproduction in Education, Society and Culture* (vol 4), London: Sage.

Bourdieu, P. and Wacquant, L.J. (1992) *An Invitation to Reflexive Sociology*, Chicago, IL: University of Chicago Press.

Bureau of Labor Statistics U.S. Department of Labor. (2016) American Time Survey Results 2016, US-17-0880. Available from: https://www.bls.gov/news.release/archives/atus_06272017.pdf

Byrnes, J.P., Miller, D.C., and Schafer, W.D. (1999) 'Gender differences in risk taking: a meta-analysis', *Psychological Bulletin*, 125(3): 367–83.

Carmi, R. et al (2011) 'Report and recommendations of the Carmi team for examining the status of women in the academic staff of higher education institutions', *Jerusalem: Israeli Council for Higher Education* (in Hebrew).

Case, S.S. and Richley, B.A. (2013) 'Gendered institutional research cultures in science: the post-doc transition for women scientists', *Community Work and Family*, 16(3): 327–49.

Ceci, S.J. and Williams, W.M. (2010) *The Mathematics of Sex: How Biology and Society Conspire to Limit Talented Women and Girls*, New York: Oxford University Press.

Chemers, M.M., Hu, L.T., and Garcia, B.F. (2001) 'Academic self-efficacy and first year college student performance and adjustment', *Journal of Educational Psychology*, 93(1): 55–64.

Chudzikowski, K. and Mayrhofer, W. (2010) 'In search of the blue flower? Grand social theories and career research: the case of Bourdieu's theory of practice', *Human Relations*, 64(1): 19–36.

Cimpian, J.R., Kim, T.H., and McDermott, Z.T. (2020) 'Understanding persistent gender gaps in STEM', *Science*, 368(6497): 1317–19.

Cochran, G., Singh, C., and Wilkin, N. (2019) *Women in Physics* (Vol 2109), 6th IUPAP International Conference on Women in Physics, New York: American Institute of Physics Publishing.

Connell, R.W. (1987) *Gender and Power: Society, the Person and Sexual Politics*, Redwood City, CA: Stanford University Press.

Connell, R.W. (1990) 'The state, gender, and sexual politics: theory and appraisal', *Theory and Society*, 19(9): 507–44.

Connell, R.W. (2019) 'New maps of struggle for gender justice: rethinking feminist research on organizations and work', *Gender, Work and Organization*, 26(1): 54–63.

Denzin, N. and Lincoln, Y. (1998) *Collecting and Interpreting Qualitative Materials*, Thousand Oaks, CA: Sage.

DeVault, M.L. (1999) *Liberating Method: Feminism and Social Research*, Philadelphia, PA: Temple University Press.

Doucet, A. (2015) 'Parental responsibilities: dilemmas of measurement and gender equality', *Journal of Marriage and Family*, 77(1): 224–42.

Eaton, A.A., Saunders, J.F., Jacobson, R.K., and West, K. (2020) 'How gender and race stereotypes impact the advancement of scholars in STEM: professors' biased evaluations of physics and biology post-doctoral candidates', *Sex Roles*, 82(3): 127–41.

Edwards, R. (1993) 'An education in interviewing: placing the researcher and the research', in C.M. Renzetti and R.M. Lee (eds) *Researching Sensitive Topics*, Newbury Park, CA: Sage, pp 181–96.

Epstein, C.F. (2004) 'Border crossings: the constraints of time norms in transgressions of gender and professional roles', in C.F. Epstein and A.L. Kalleberg (eds) *Fighting for Time: Shifting Boundaries of Work and Social Life*, New York: Russell Sage Foundation, pp 317–40.

Eran Jona, M. and Nir, Y. (2019) 'Women in physics in Israel', in G. Cochran, C. Singh, and N. Wilkin (eds) *Women in Physics* (Vol 2109), 6th IUPAP International Conference on Women in Physics, New York: American Institute of Physics Publishing.

Eran Jona, M. and Nir, Y. (2020) 'Ph.D. in physics as a hurdle race, and the "glass hurdles" for women', arXiv preprint, Available from: https://arxiv.org/abs/2007.02251

Eran Jona, M. and Perez G., (2020) 'What can the 'Start-Up Nation' do to enhance diversity in physics?', *Weizmann Institute of Science research report*.

Falk, N.A., Rottinghaus, P.J., Casanova, T.N., Borgen, F.H., and Betz, N.E. (2017) 'Expanding women's participation in STEM: insights from parallel measures of self-efficacy and interests', *Journal of Career Assessment*, 25(4): 571–84. DOI: 10.1177/1069072716665822

Fogiel-Bijaoui, S. (1999) 'Families in Israel: between familism and post-modernism', in D. Izraeli, A. Freidman, H. Dahan-Kalev, S. Fogiel-Bijaoui, H. Herzog, M. Hasan, and H. Naveh (eds) *Sex, Gender, Politics: Women in Israel*, Tel-Aviv: Hakibbutz Hameuchad Publishing House, pp 107–66 (in Hebrew).

Gerson, J.M. and Peiss, K. (1985) 'Boundaries, negotiation, consciousness: reconceptualizing gender relations', *Social Problems*, 32(4): 317–31.

Glesne, C. and Peshkin, A. (1992) *Becoming Qualitative Researchers: An Introduction*, New York: Longman.

Gofen, A. (2011), 'Academic career of graduates with excellent grades in SET fields 1995–2005', *Research Report of the Federman School of Public Policy and Government*, The Hebrew University of Jerusalem (in Hebrew).

Gonsalves, A.J., Danielsson, A., and Pettersson, H. (2016) 'Masculinities and experimental practices in physics: the view from three case studies', *Physical Review Physics Education Research*, 12(2): 020120.

Goulden, M., Frasch, K., and Mason, M.A. (2009) *Staying Competitive: Patching America's Leaky Pipeline in the Sciences*, Berkeley, CA: Center for American Progress.

Grunert, M.L. and Bodner G.M. (2011) 'Finding fulfillment: women's self-efficacy beliefs and career choices in chemistry', *Chemistry Education Research and Practice*, 12: 420–6.

Guy, A. and Shnider, A. (2019) 'Old characteristics of gender discrimination in the new workforce structure', The Heth Academic Center for Research of Competition and Regulation publication (in Hebrew).

Gvozdanovic, J. and Maes K. (2018) 'Implicit bias in academia: a challenge to the meritocratic principle and to women's careers – and what to do about it?', retrieved from *League of European Research Universities*, Available from: www.leru.org/publications/implicit-bias-in-academia-a-challenge-to-the-meritocratic-principle-and-to-womens-careers-and-what-to-do-about-it

Hartsock, N. (1987) 'Rethinking modernism: minority vs. majority theories', *Cultural Critique*, 7: 187–206.

Heilman, M.E. and Haynes, M.C. (2008) 'Subjectivity in the appraisal process: a facilitator of gender bias in work settings', in E. Borgida and S.T. Fiske (eds) *Beyond Common Sense: Psychological Science in the Court-Room*, London: Blackwell, pp 127–55.

Hodkinson P. and Sparkes, A.C. (1997) 'Careership: a sociological theory of career decision making', *British Journal of Sociology of Education*, 18(1): 29–44.

Holvino, E. (2010), 'Intersections: the simultaneity of race, gender and class in organization studies', *Gender, Work and Organization*, 17(3): 248–77.

Ivie, R., White, S., and Chu, R.Y. (2016) 'Women's and men's career choices in astronomy and astrophysics', *Physical Review Physics Education Research*, 12(2): 020109.

Johnson, R.B., Onwuegbuzie, A.J., and Turner, L.A. (2007) 'Toward a definition of mixed methods research', *Journal of Mixed Methods Research*, 1(2): 112–33.

Johnson, P.A., Widnall, S.E., and Benya, F.F. (eds) (2018) *Sexual Harassment of Women: Climate, Culture, and in Academic Sciences, Engineering, and Medicine*, National Academies of Sciences, Engineering, and Medicine, Washington, DC: National Academies Press. DOI: 10.17226/24994

Kaatz, A., Gutierrez, B., and Carnes, M. (2014) 'Threats to objectivity in peer review: the case of gender', *Trends in Pharmacological Sciences*, 35(8): 371–3.

Kanny, A.M., Sax, L.J., and Riggers-Piehl, T.A. (2014) 'Investigating forty years of STEM research: how explanations for the gender gap have evolved over time', *Journal of Women and Minorities in Science and Engineering*, 20(2): 127–48.

Kanter, R.M. (1977) *Men and Women of the Corporation*, New York: Basic Books.

Kark, R. (2007) 'Women in the land of milk, honey and high technology: the Israeli case, in women and minorities', in R.J. Burke and M.C. Mattis (eds) *Science, Technology, Engineering and Mathematics*, Cheltenham: Edward Elgar, pp 101–27.

Komter, A. (1991) 'Gender power and feminist theory', in K. Davis, M. Leijenaar, and J. Oldersma (eds) *The Gender of Power*, London: Sage, pp 420–62.

Krumer-Nevo, M., Lavie-Ajayi, M., and Hacker, D. (eds) (2014) *Feminist Research Methodologies*, Tel-Aviv: Hakibutz Hameuchad Publications, Migdarim Series (in Hebrew).

Kulis, S., Sicotte, D., and Collins, S. (2002) 'More than a pipeline problem: labor supply constraints and gender stratification across academic science disciplines', *Research in Higher Education*, 43(6): 657–91.

Lamont, M. and Molnár, V. (2002) 'The study of boundaries in the social sciences', *Annual Review of Sociology*, 28(1): 167–95.

Lent, R.W., Lopez, F.G., and Bieschke, K.J. (1991) 'Mathematics self-efficacy: sources and relation to science-based career choice', *Journal of Counseling Psychology*, 38(4): 424–30.

Lent, R.W., Brown, S.D., and Hackett, G. (1994) 'Toward a unifying social cognitive theory of career and academic interest, choice, and performance', *Journal of Vocational Behavior*, 45(1): 79–122.

Lewis, K.L., Stout, J.G., Pollock, S.J., Finkelstein, N.D., and Ito, T.A. (2016) 'Fitting in or opting out: a review of key social-psychological factors influencing a sense of belonging for women in physics', *Physical Review Physics Education Research*, 12(2): 020110.

Long, J.S. (ed) (2001) *From Scarcity to Visibility: Gender Differences in the Careers of Doctoral Scientists and Engineers*, National Research Council, Washington, DC: The National Academies Press. DOI: 10.17226/5363

Mizrahi, S. (2015) 'Women in Israel: central issues', the Knesset Research and Information Center report (in Hebrew).

Mulvey, P. and Pold, J. (2020) 'Physics doctorates: skills used and satisfaction with employment', data from the Degree Recipient Follow-Up Survey for the Classes of 2015 and 2016, Focus On, AIP Statistical Research Center.

Nielsen, M.W. (2017) 'Reasons for leaving the academy: a case study on the "opt out" phenomenon among younger female researchers', *Gender, Work and Organization*, 24 (2): 134–55. DOI: 10.1111/gwao.12151

Porter, A.M. and Ivie, R. (2019) *Women in Physics and Astronomy*, technical report, AIP Statistical Research Center.

Reinharz, S. and Davidman, L. (1992) *Feminist Methods in Social Research*, New York: Oxford University Press.

Reskin, B.F. and Roos P.A. (1987) 'Status hierarchies and sex segregation', in C. Boss and G. Spitze (eds) *Ingredients for Women's Employment Policy*, Albany, NY: SUNY Press, pp 3–21.

Scott, J.W. (1986) 'Gender: a useful category of historical analysis', *American Historical Review*, 91(5): 1053–75.

Seierstad, C. and Kirton, G. (2015) 'Having it all? Women in high commitment careers and work–life balance in Norway', *Gender, Work and Organization*, 22(4): 390–404. DOI: 10.1111/gwao.12099

Sheppard, D.L. (1989) 'Organization power and sexuality: the image and self-image of women managers', in J. Hearn, D.L. Sheppard, P. Tancred-Sheriff, and G. Burrell (eds) *The Sexuality of Organization*, London: Sage, pp 139–57.

The Israeli Central Bureau of Statistics (ICBS) (2020) 'Fertility rates of Jewish women and other in Israel by level of religiosity 1979–2017', Table No. 1, Available from: www.cbs.gov.il/EN/pages/ default.aspx

Toren, N. (2009) 'Mizrahi women in academia: gender, ethnicity and social status', in R. Herz-Lazarowitz and I. Oplatka (eds) *Gender and Ethnicity in the Israeli Academy*, Haifa: Pardes Publications (in Hebrew), pp 70–88.

Traweek, S. (1988) *Beamtimes and Lifetimes: The World of High Energy Physicists*, Cambridge, MA: Harvard University Press.

Van Maanen, J. and Schein, E.H. (1979) 'Toward a theory of organizational socialization', in B.M. Staw (ed) *Research in Organizational Behavior*, Greenwich, CT: JAI, pp 209–64.

Weitz, R. (2001) 'Women and their hair: seeking power through resistance and accommodation', *Gender and Society*, 15(5): 667–86.

Wolfinger, N.H., Mason, M.A, and Goulden, M. (2008) 'Problems in the pipeline: gender, marriage, and fertility in the ivory tower', *Journal of Higher Education*, 79(4): 388–405.

Yair, G. (2020) 'A different reason: how Israeli scientists think about careers and family life', *Israel Studies*, 25(2): 159–78.

5

Difficult to Recognize But Harmful: Experiences of Microaggressions among European Women Physicists

Paulina Sekuła

Introduction

Women are under-represented among researchers and academic teachers in the disciplines of science, technology, engineering, and mathematics (STEM) in Europe and worldwide (Guillopé and Roy, 2020; European Commission, 2021). Physics belongs to these STEM fields where gender imbalance is particularly pronounced (Elsevier, 2017). Among the factors that are conducive to the prevalence of gender disparities, the roles of organizational culture and climate are widely discussed (Benschop and Brouns, 2003; O'Connor, 2020). There is substantial evidence that women are systematically treated differently in academia and science disciplines, including physics. While there is also some competing evidence that overt sexism is nowadays relatively insignificant (Ceci and Williams, 2010; Hughes, 2014), covert discrimination – that which is based on subtle, often unconscious gender bias – prevails in academia. It includes gender microaggressions: subtle insults taking place during everyday interactions that communicate a hostile or dismissive attitude towards women (Sue, 2010; Yang and Carroll, 2018). They might be perceived as harmless elements of workplace culture. However, their cumulative nature creates a hostile and invalidating climate for women and may have detrimental effects on their psychological well-being, self-confidence, comfort at work, and productivity (Simatele, 2018; Periyakoil et al, 2019; Blithe, 2020).

The aim of this chapter is to examine gendered experiences of women in physics through the analytical lens of microaggressions. It identifies common forms of subtle discrimination faced at work by European women physicists, explores the strategies women use to cope with the unfriendly climates of their workplaces, and investigates the reported consequences of being exposed to microaggressions. While the problem of gender discrimination in physics might be undervalued – as some physicists – both men and women – still perceive it as non-existent – detrimental effects of microaggressions are evident. While this has been explored in few empirical studies on women physicists working in US universities, this chapter fills the lacuna in research on other regions. The analysis is based on a qualitative study performed under the framework of the H2020 project Gender Equality Network in the European Research Area (GENERA). It covers the results of 40 semi-standardized interviews conducted in 2017 with women physicists working in 12 European research-performing organizations and higher education institutions (for a wider discussion of the GENERA project, see Chapter 7 in this book).

Gender imbalances in physics

Although the proportion of women among physics students and researchers is increasing in some regions of the world, the gender disparity remains significant and widens with the advancement of scientific careers (Elsevier, 2017; de Hoogh et al, 2019; Porter and Ivie, 2019). The Elsevier (2017) report indicates that between 2011 and 2015, women accounted for about 25 per cent of EU-28 researchers in physics and astronomy, which, while representing an increase of 8 percentage points compared with 1996–2000, still meant that the under-representation of women in these disciplines is among the highest. In individual countries included in this report, women's participation in physics ranged from 11 per cent in Japan to 38 per cent in Portugal (Elsevier, 2017). In addition, non-aggregated data from other sources and for selected countries and research units suggest that women are sometimes absent from professorial or equivalent positions altogether, and that it is rare to have a share of women physics professors above 30 per cent (see Antolini et al, 2019; Eran Jona and Nir, 2021; Miikkulainen et al, 2019; Porter and Ivie, 2019; Šatkovskiene et al, 2019; Stojanović et al, 2019). Holman and collaborators (2018), analysing the rate of increase in the proportion of women among authors of scientific articles, argue that in physics – and disciplines such as mathematics, computer science, and surgery – the abolition of gender disparity will not occur within this century given present-day rates of change.

Gender segregation in physics also includes low women's representation among those in leadership positions. While this information is lacking at an aggregate level, partial reports illustrate the extent of gender disparity among managers of physics research organizations. For example, the proportion of women among those heading the 30 physics departments in the US and the 20 largest physics departments in the world did not exceed 10 per cent in 2017 (McCullough, 2019).

The determinants of the under-representation of women in science – including physics – constitute a network of multiple interrelated factors (Charles and Bradley, 2009; Ceci et al, 2014; O'Connor et al, 2015) of both formal and informal nature (De Welde and Laursen, 2011), interacting at individual, organizational, and systemic levels (Timmers et al, 2010; O'Connor et al, 2015; Zippel, 2017; Savigny, 2019; O'Connor, 2020; see also Chapter 6 in this book). Among them the role of organizational culture and climate is widely discussed (Benschop and Brouns, 2003; O'Connor, 2020). The data confirm that on average, women are more likely than men to opt out from STEM fields, because they 'do not feel as they fit and are accepted' (Lewis et al, 2017: 8). Among the key elements of the chilly climate for women in STEM fields are microaggressions.

What are microaggressions?

Microaggressions are verbal or non-verbal exchanges that communicate hostile, derogatory, or negative slights, invalidations, and insults to an individual or group because of their marginalized status in society. They can span a continuum from being conscious and deliberate to unconscious and unintentional (Sue, 2010). Among various forms of microaggressions there are microassaults, microinsults, or microinvalidations. Microassaults are closest to overt forms of discrimination as they are often conscious, explicit derogations characterized by violent verbal or non-verbal attacks meant to assail the recipient's identity through name-calling, avoidant behaviour, or purposefully discriminatory behaviours. Both microinsults and microinvalidations are forms of microaggressions that are usually performed unconsciously. Microinsults refer to communications that convey rudeness and insensitivity and demean a person's heritage, such as messages that certain groups are less competent. Microinvalidations relate to communications that exclude, negate, or nullify thoughts, feelings, or experiential reality of a recipient; for example, a denial of existence of discrimination (Sue and Spanierman, 2020).

Microaggressions happen to occur in patterned ways to 'remind marginalised people of their vulnerable position in social hierarchies' (McClure and Rini, 2020: 6); however, they are often difficult to recognize due to their subtlety and ambiguous character (Friedlaender, 2018). Being pervasive and automatic in everyday interactions, their impact is easily trivialized and minimized. However, their cumulative nature creates a hostile and invalidating climate for the recipients and may have detrimental effects on their standard of living and working (Sue, 2010).

While early theorizing of microaggressions referred to the analysis of discriminatory practices against people of colour in American society, current debates concentrate on microaggressions expressed towards any marginalized groups, based on such characteristics as ethnicity, gender, sexual orientation, age, or socio-economic status (Berk, 2017; Sue and Spanierman, 2020). Although most studies refer to the US context, there is some research on microaggressions conducted in other regions, including Europe (Estacio and Saidy-Khan, 2014; Piccinelli, 2020). The occurrence of gender microaggressions has been identified in the context of higher education and research (Barthelemy et al, 2016; Gaisch et al, 2016; Blithe and Elliot, 2020) and include instances of misrecognition of women's professional abilities, devaluing their achievements, infantilizing or stereotyping them, objectifying them as sex objects, and treating them in such a manner as to deny them equal access and opportunity (Sue and Spanierman, 2020). It is argued that they 'act upon women in several ways, by reiterating the social view that men are more valued than women, by reinforcing traditional stereotypes about proper gender roles, and by contributing to violence toward women by objectifying and sexualizing them' (Barthelemy et al, 2016: 4).

Gender microaggressions in physics: behavioural manifestations of implicit bias

Physics is a masculine field characterized not only by numerical over-representation of men, but also by a masculine culture and the public image of the field (Sekula et al, 2018; Krzaklewska et al, 2019; Eran Jona and Nir, 2021). While physicists portray the world of science in terms of a 'culture of no culture', that is completely devoid of the influence of human factors, including gender (Traweek, 1988: 162), there operate specific stereotypes of people in the field and negative stereotypes of women's abilities (Carli et al, 2016; Cheryan et al, 2017; see also Brage and Drew, Chapter 2 in this book). The presence of women in physics remains an 'anomaly' (Keller, 2001), which is explained in terms of the underlying associations between masculinity and the mathematical nature of theory in physics and the technological nature of experimental practice (Traweek, 1988; Cronin and Roger, 1999; Rolin and Vainio, 2011).[1] The practice of physics is therefore gendered, characterized by certain styles of doing science that are linked to certain ideals of masculinity, including the creative 'happy boy' and the 'priest' engaged in the most fundamental field among the sciences (cited in Rolin 2008; Rolin and Vainio, 2011). Physics laboratories are particular spaces for the enactment of masculinities encompassing qualities such as physical fitness, machine literacy, or creativity in their operation; qualities that women are believed to be lacking (Gonsalves et al, 2016).

Bias towards women and gender stereotypes may translate into practices of gender discrimination. Overt gender discrimination, albeit less pronounced than in earlier times (Ceci and Williams, 2010; Hughes, 2014), continues to be identified in academia (Blithe and Elliot, 2020), including physics (Aycock et al, 2019). However, subtle discriminatory practices seem to be widespread. The occurrence of microaggressions in the field of physics – and other STEM fields – has been reported in a few qualitative (Barthelemy et al, 2014, 2016; Ong et al, 2018) and quantitative studies (Yang and Carroll, 2018; McCullough, 2020) carried out in the US context. Women physicists report 'experiencing small actions that challenge their identity as scientists, denigrate their abilities or make them feel like outsiders on a regular basis' (Harris, 2016: 46). Barthelemy et al (2014) found evidence that undergraduate and graduate physics women students encounter microassaults, microinsults, and microinvalidations. First, microassaults – which are purposeful, conscious acts – the women experienced as sexist jokes, references to their inferiority, and restrictions from laboratory equipment. Microinsults – which are more subtle, and often unconscious – include ignoring and overlooking women in everyday interactions, convey a message that they are not competent, do not belong to the physics community, and that their ideas are not important. Finally, microinvalidations are manifested in lack of reaction when women report cases of gender discrimination, suggesting that their experiences are non-existent or not harmful. In another study, the authors (Barthelemy et al, 2016) used the typology of Sue (2010) to identify eight sub-themes of microaggressions experienced by women physicists. 'Sexual objectification' reduces women to their physical appearance or implies that their bodies should be controlled and commodified by men. 'Second-class citizenship' manifests in a belief that women are less as a group and should not have the same access to resources and opportunities as men. Similarly, 'sexist language' infers the superiority of men over women. 'Assumption of inferiority' carries ideas of women's less intellectual or physical capabilities. 'Restrictive gender roles' manifest in a belief that women should play traditional feminine roles. 'Denial of the reality of sexism' means not believing that sexism exists. 'Invisibility' implies overlooking, not including or recognizing women and their talent, abilities, or input in the workplace or in a wider society. Finally, 'sexist humour or jokes' include crude jokes about women, rape, and domestic violence that are demeaning, filled with negative stereotypes about women and reinforce restrictions on their behaviour.

Microaggressions might be perceived as harmless elements of workplace culture and treated as only 'the normal sounds of human rambling' (Etzioni, 2014). However, as they are repetitive, ubiquitous, and often occur in the context of formal power or informal domination, microaggressions can have detrimental consequences for women's psychological well-being, self-confidence, comfort at work, and productivity (Simatele, 2018; Miner et al, 2019; Periyakoil et al, 2019; Blithe and Elliot, 2020), and, therefore, their professional development and advancement. Accumulation of microaggressions, together with more explicit, overt forms of gender discrimination, makes the climate for women 'chilly' and sends them the message that physics is a man's world (Ivie and Ray, 2005; Hughes, 2014) and lowers women's sense of belonging to the field (Lewis et al, 2017).

In the European context, the perspective of microaggressions has not been widely used in the analysis of discriminatory practices in research and academia. While subtle forms of gender discrimination are as well investigated and reported, they are, however, differently conceptualized. Among these conceptualizations there is a concept of non-events used to analyse sexism in a Finnish context (Husu, 2013, 2020). Non-events are things and actions that do not happen to women academics but would have been desired for their career development and for their well-being at work. Different forms of non-events include ignoring or bypassing women's research and performance; not citing their work; not inviting or welcoming them to important informal and formal networks; bypassing them for awards; not giving them merit-advancing tasks, such as representing the research group in public forums; not asking them to design or participate in scientific meetings, conferences, panels, or as keynote speakers; or staying silent when it comes to career support, advice, and mentoring (Husu, 2013). It is worth noting that non-events are here analogous to previously described invisibility as one of the gender-microaggressive themes. Just as in the case of microaggressions, non-events are also difficult to recognize and respond to, but their accumulation over time can have a negative impact on women's careers and psychological well-being.

Women employ individual strategies to survive in a hostile environment of masculinized science, where they are treated as outsiders (Fotaki, 2013; Haas et al, 2016; O'Connor et al, 2018). To avoid isolation, they often develop a strategy of assimilation to the masculine norms and denial of gender discrimination, and try to blend in, to be 'one of the boys' by focusing on achievements, increasing individual effort and reducing private commitments (Rolin and Vainio, 2011; Haas et al, 2016; O'Connor et al, 2018).

Study methodology

The data used in this chapter stems from the H2020 project GENERA aiming at designing and implementing interventions contributing to overcoming the under-representation of women in physics and opening up more opportunities for women to create successful careers in physics research and in related fields in European higher education institutions and research institutes.[2] To explore career paths of European physicists, explain gender inequalities in physics, and identify strategies to overcome barriers that hinder the development of women physicists' careers, the method of semi-standardized interviews was used. It is based on the assumption that people as social actors construct their 'subjective theories' about their life and experiences. The notion of 'subjective theory' refers to the fact that 'the interviewees have a complex stock of knowledge about the topic under study' (Flick, 2006: 155). In addition to the semi-standardized interviews, expert interviews with physicists occupying leading positions were conducted focusing on the perception of gender inequalities in the researched organizations as well as the evaluation of existing and possible measures to counteract them. Altogether 83 interviews were conducted in 2017 with women and men physicists working in 12 institutions (both research institutes and universities) operating in eight European countries. The results from this study have been discussed in the project report (Sekula et al, 2018).

While the research sample for the project study covered the perspectives, experiences, and standpoints of both women and men physicists, both researchers and organizational leaders, this analysis includes only the results of the interviews with women researchers (see Table 5.1).[3] The research sample (40 interviews) is diversified to include both young researchers at the start of their career path (PhD students, postdocs, research assistants) and senior researchers who hold independent positions (researchers, senior researchers, full professors), working in different sub-disciplines of both theoretical and experimental physics: from optical physics, through astrophysics and cosmology to nuclear physics and biophysics. Within the sample 19 respondents have children (mostly one, sometimes two, of diverse ages, also adult ones).

Table 5.1: Characteristics of the interviewees

Country	Career stage		Children	
	Early	Senior	No	Yes
Germany – 3 institutions	6	2	7	1
Netherlands	1	0	1	0
Italy – 2 institutions	3	4	5	2
Spain	2	2	1	3
Switzerland	1	1	0	2
Romania	1	3	1	3
France	2	1	2	1
Poland – 2 institutions	5	6	5	6
Total number of interviews	21	19	22	18

The interview guidelines were prepared in English by the team of sociologists from the Jagiellonian University in Kraków, in which the author of this chapter participated. The guidelines were translated into national languages of the participating organizations and applied with necessary adjustments to local contexts.[4] Interviews were conducted by local researchers in the languages of the country in question or in English. For each interview, a structured note with the summary of interview themes and selected citations was prepared in English by local researchers. In the next step, all structured notes were examined with the content-analysis technique. The interview data were analysed within an iterative process of assigning specific data-driven and theory-driven codes, their verification, and integration in more general categories in order to avoid repetitions. For this analysis the codes were developed from the concept of the microaggressions and relevant research literature.

Forms of microaggressions, their consequences, and the ways of coping with them have been identified on the basis of both the incidents reported freely in the interviewees' narratives on their career paths in physics and the working cultures of their organizations, as well as the answers to a direct, open-ended question on the experiences of being unequally treated or discriminated.[5] With this approach we were able not only to reach the stories of gender microaggressions which interviewees deemed salient enough to recall them without being directly asked, but also to 'dig deeper' into their memory and possibly discover previously neglected incidents.

The limitations of presented research result from different sizes of organizational research samples and their varying internal diversification in respect to career stage, cultural differences in understanding the concepts and notions in questions, and the variable quality of structured notes prepared by the local researchers. Due to these limitations, no country or organizational comparisons of results are applied. Specifying countries in which the respondents were employed, together with their academic positions, was omitted in the interest of anonymity.

Manifestations and impact of gender microaggressions in physics

When directly asked about being discriminated against, 19 women physicists denied they had ever been treated differently because of their gender. However, 5 out of these 19 women report further in the interview incidents that can be easily interpreted as microaggressions. In sum, various forms of gender-microaggressive experiences have been identified in the interviews with 23 women researchers being both at the beginning of their career and at a more advanced level.[6] Before moving to the analysis, the interviewees' general perception of gender inequalities in physics is here outlined.

While quite a few of the examined women physicists present rather ambivalent attitudes towards the issue of gender inequalities in physics, in the narratives of one third of interviewees (14) there appears conviction that gender discrimination remains a living problem in their institutions and in the wider environment of physics. Some of them only signal a problem by talking about "existing discrimination" (46_F), or "misogynous men" (40_F), or women being a "little excluded" (06_F). Others, more precisely, talk about the climate of physics, which includes believing "that physics is more technical than other research fields and can be difficult for a woman to handle with" (83_F), looking down on young women, sexualizing them, and overburdening them with teaching duties as well as mobbing and sexual harassment. A few interviewees declare awareness of the presence of gender discrimination in their environments even despite the lack of own experience of unequal treatment. While one of the physicists explicitly defines her situation as being exceptional, another links her advantage of not being discriminated against to the qualifications of her (male) colleagues:

'The question is asked to me very often, so it really forced me to think about it and remember. I happily never had this feeling. I want to say it's probably an exception; I have many friends who experienced this kind of discrimination in personal or professional areas.' (36_F)

'My enlightened colleagues never had problems with the fact that I am a woman and I have never felt any problem connected with this. ... but I know there are women ... who feel that [way].' (56_F)

On the other hand, 12 interviewees – basing their reflections on their own experience – explicitly deny any existence of gender discrimination in physics. They argue that, while it might have been a systemic problem in the past, nowadays – even if it happens – it is rather rare and diminishing or happens elsewhere outside physics; that is, in other scientific disciplines or in the corporate world. To validate their perceived lack of discrimination in the field, they claim that physics is "fair and equal", there is no prejudice, and everyone is treated similarly. At the same time, they explain gender disproportions as the result of gender differences in life choices and natural predispositions (see Krzaklewska et al, 2019). In this context it is worth mentioning that denial of the existence of sexism can be one of the individual coping strategies that have been identified to be employed by women working in high-ranking positions or traditionally male-dominated fields (such as physics), which allows them to perceive the system as fair and, therefore, enhance their subjective well-being (Napier et al, 2020) and to assimilate to the masculine world (Haas et al, 2016) or, in other words, to blend in (O'Connor et al, 2018).

The themes of microaggressions that have been identified in the narratives of women interviewees are discussed here in the order of frequency of occurrence.

Presumed incompetence

As previously, we discussed conceptualizations of two separate themes of gender microaggressions – the assumption of inferiority and second-class citizenship (Sue, 2010; Barthelemy et al, 2016). Whereas an assumption of inferiority implies the expression of the conviction that women are inherently unable to do certain tasks due to their physical or intellectual inferiority, particularly in comparison to men; second-class citizenship refers to treating women as lesser persons or group and/or believing that women should not have the same access to resources and opportunities as men (Barthelemy et al, 2016). While there is analytical distinction between these two categories, in the analysed narratives of women physicists they are difficult to differentiate, as they interrelate and both explicitly refer to cases when women are told through words and/or actions that they are less competent than men. Therefore, these two themes are analysed here together under the heading of 'Presumed incompetence' (see Gartner, 2021).

The theme of presumed incompetence appears in the narratives of 11 women physicists at different career stages. They recall situations taking place during their studies or at work when they were 'informed' with words or behaviour by their male supervisors, directors, or colleagues of their inabilities. These included being an individual or collective addressee of microassaultive statements and demeaning labels of "not having spatial imagination", "not being able to worthily represent the department", "being second-class researchers", "being of a second, third or C category", "having again stuffed something up", or "being stupid" as a woman. According to one of the respondents this kind of discrediting of one's abilities does not happen among women scientists:

'And for example with women, something like this does not happen. So, it never happened to me, that I tried to talk to a woman and she immediately said, immediately looked at me strangely or immediately said "Ugh, what are you doing here?" or "Why are you coming to me? You are not able to do anything" or something.' (07_F)

Some of the women described situations when their profession or research field evoked reactions of astonishment and disbelief. They interpret comments like "oh, really math and physics?" or receiving emails addressed to them as if they were men as the effect of the domination of an opinion that a physics is a manly thing ("A genius has to be a man"). These are rather indirect and unconscious forms of microaggressions:

'I felt that he [IT Programming course coordinator] didn't know how to address the fact that I was a woman doing such an abstract thing [theoretical physics – P.S.].' (51_F)

'I already heard lots of times "That's nice you got this job because you are a girl". People don't realize what they are actually saying, that means "I don't think you are qualified enough for this job".' (37_F)

Women physicists recall also other particular verbal and behavioural indignities directed towards them throughout their career which either informed them that expectations towards them are lower or were aimed at showing them their proper place in the hierarchy. These actions include undermining their abilities and competence to make accurate measurements, fix laboratory equipment, or write good articles:

'We had this experiment and there were very vulnerable electronic modules, they were very capricious and had to be monitored all the time. It was my experiment, so I asked my male colleagues "If something happens to it, let me know, please?" and then they asked me a question "And what will you do? Which expert should we call? What will you do?" and I said that I would come and repair it because I knew how. But for them it was impossible to imagine that a girl could do such things. Although they are educated men and probably have seen many women doing different things, yet still the environment is strongly dominated by men, and they couldn't imagine me doing such things. All in all, they are not against women, but they have in their minds these clichés that we [women] don't do such things.' (61_F)

The assumption of inferiority or incompetence also emerges in some experiences of not being accepted in leadership positions, be it a chairperson of a students' association, or a head of a department. The following two narratives illustrate the persistence of the belief that women lack leadership competences, and that power is masculine:

'One year [during MSc studies] I ran for the president of the [students' scientific] association and first there was an argument that they didn't want someone from the third year to be the president, and then they elected my colleague from the year, so here I am sure it was because the guys didn't want to be ruled by a chick.' (55_F)

'I was the best among them, as far as scientific output, the level of engagement and the number of promoted doctors are concerned. Still, men thought that it was one of them who should be the head of the department, which is why they moved to another department when I took over this position.' (70_F)

The physicist sharing the experience of not being accepted as a departmental leader also points out that being a woman, she had been discouraged from applying for promotion:

'at the last stage of the professorship [procedure – P.S.], I found that there were doubts here "because you're too young to be a professor" or "wait yet", ... and certainly this kind of question would not have happened to a guy, it would have been more to say to him, "well, why are you not running for professor yet?"' (70_F)

Restrictive gender roles

The theme of restrictive gender roles refers to the belief that women must uphold traditional gender roles. The message which is communicated with this microaggression is that women either do not belong in the field of physics, as their presence there is in direct conflict with their social roles as women (Barthelemy et al, 2016), mainly as caregivers and support-providers, or that, since they are physicists, they should take tasks which are 'appropriate' for them. At least 11 women respondents discuss cases of being the object of this kind of microaggression.

A few respondents reflect on having heard during their studies that women are not fit for physics as they "are meant to play with dolls" and do the housework. While these microassaultive statements directly refer to traditional division of gender roles to legitimize negative attitudes towards the presence of women in physics, they also imply an assumption of their inferiority in comparison with men.

Restrictive gender roles manifest further with cases where women are not recognized as physicists. One of the women respondents recalls a situation when she was mistaken for a secretary: "... they thought I was gonna go and make them a coffee, because I was a woman. They thought I was a secretary and not a scientist" (10_F).

Women also report being reminded that they should not challenge the traditional division of gender roles when they become mothers. The status of motherhood is seen by others as a self-explanatory reason for – probably mostly unintentionally – limiting women their career opportunities: "When my child was born, I received an email saying that my invited talk had been replaced by a poster. And I said 'no, you cannot do this' and finally I was allowed to give the talk by Skype" (26_F).

Being a scientist with children becomes a heavily gendered experience, especially since men are not expected to sacrifice their career when they become parents:

'The organizer heard I gave good seminars so he invited me but he told me: "I know that you have a kid so I was very hesitant to invite you. I didn't want to put you in a difficult situation." You would never say that to a guy that has a kid, right?' (05_F)

'The clash between social expectations about proper gender roles on the one hand, and the image of physics as requiring full-time engagement on the other, becomes an exclusionary practice and a dead-end trap for some women. They are both expected to limit their professional activity when having babies and, at the same time, they are told that they "do not have good qualifications for being a physicist, because they are too much distracted by home, by children".' (65_F)

Once accepted as physicists, women are often still expected to play their 'natural', socially acceptable roles. Women physicists share their experiences of being automatically assigned to service duties such as taking notes at committee meetings, making copies of documents, and making coffee before and washing dishes after the scientific group's meetings. Similarly, administrative tasks are sometimes ascribed by default to women: "If there were any administrative duties my director at my previous institute wanted us, as women, to do these things. I have always refused doing this kind of thing. I don't know why women must always do administrative tasks" (40_F).

Some women respondents working in higher education institutions also reckon that their role as physicists happens to be reduced to teaching students. While all these previously mentioned duties that are reported to be often passed to women are necessary for the efficient functioning of an institution or scientific teams, they are time-consuming, less valued, and carry less weight in promotion procedures: "I heard it already twice that they like me as a teaching assistant. That is nice, but I never heard my supervisor say that. She knows that there is always more work than that to be done" (51_F).

Invisibility

Invisibility refers to the cases of not including or recognizing women within and outside their institution, and/or not being heard or listened to by their peers due to their gender (Sue, 2010; Barthelemy et al, 2016). This theme also fits well into the category of non-events that have been discussed previously. In the interview sample, nine women physicists reported having these microinsultive experiences.

Invisibility of women in the workplace manifests itself in being ignored, overlooked, and not being asked to conferences or meetings, which makes them feel that they "don't belong to the community" (06_F).

'And then a man comes over and we say "Yes, hello, we are blah and blah" and this guy, this guy, he only greets the man and I don't matter at all. Just straight up ignored, from beginning to end. Completely. Despite the fact that my colleagues sometimes said "Yes, she is well versed, maybe she can say something about it" or something, they always went back to the other man.' (07_F)

Not only women experience not being included in conversations but also their ideas are sometimes underestimated, they "are not taken seriously at all" (07_F), or are not given "the value they deserved" (26_F):

'There were things which happened to me where I thought, "that's because I'm a woman", but I was never sure. Like when I was a postdoc, I had an idea, and nobody was interested. It was another postdoc but this time a male and people were interested. It made me think if I had been a man they might have listened.' (01_F)

Another way of making women invisible is ignoring their performance by attributing merit and success to their male colleagues. This kind of behaviour can be seen as an aspect of a wider, well-identified phenomenon of systemic under-recognition of women's scientific efforts and achievements, labelled as 'the Matilda effect' (Rossiter, 1993; Wennerås and Wold, 1997; Benschop and Brouns, 2003):

'I was then, how to say it, an organic adaptor, who could insert a screw in such a way that it was possible to take measurements. So I sat there whole time and was doing it. No one else was able to regulate gas flow in such a way that the temperature stayed static. And then it was my male colleague who got all the praise, he wrote a paper, because he analysed the measurements. So, what that he tried five times to take measurements on his own, but he couldn't.' (65_F)

Women also report cases of their superiors staying silent when it comes to career support, including financing participation in an expensive scientific summit or assigning a project:

'Yesterday I talked to the other woman and then it came out that we are both sort of isolated. So we are not allowed to now, we both do not have a project anymore, so no defined project and they [the men] are getting all the new projects, even though one of them is leaving in two months.' (07_F)

Sexist jokes and comments

Nine women respondents refer in different parts of their interviews to sexist jokes or comments that were told in their presence by men. Mostly without citing them, the interviewees call these jokes "inappropriate", "stupid", "insinuating", "boorish", "chauvinist", "distasteful", or "obscene". The narratives of some interviewees suggest that telling sexist jokes is a part of academic culture that women physicists are faced with on a regular basis both during conversations with co-workers as well as at meetings with their superiors: "A few times my boss did some insinuating jokes about the fact that I share the office with a male colleague" (83_F). "When the director welcomed new staff he made a joke, that made me uncomfortable" (51_F).

Other respondents asses that occurrence of this form of microaggression is becoming an outdated problem, as they happened rather more in the past than nowadays and were used mainly by elderly professors:

'Mainly elderly people, ... older than 50 years, used such sexual overtones or jokes, but it slowly, it somehow slowly changes, however there used to be such distasteful comments.' (64_F)

'The conviction that telling sexist jokes and making sexist remarks has ceased to be an accepted norm is further supported by an observation that even if they take place, they are met with disapproval: "he said it as a great joke, but nobody laughed actually".' (61_F)

Sexual objectification

Sexual objectification means treating women as sexual objects, reducing them to their physical appearance or assuming their bodies should be controlled and commodified by men (Barthelemy et al, 2016). Experiences described by six women researchers can be classified as this kind of microaggression. These women report being treated by their superiors and male colleagues as physical, sexual objects:

"It happened once, as I am very sensitive to this kind of thing, so maybe it was because of my sensitiveness. Once, it was at the first year of studies, we came to one professor and I carelessly, I never wear short skirts, had a dress long to the knee, and he was watching me carefully, evidently, like three times." (56_F)

Women physicists also evoke cases of being flirted with at work or during conferences and – as one of the interviewees put it – approached "with weird sexual ... ulterior motives" (07_F), which are difficult to resist, as this may have an unintended impact in the form of retaliation, especially when the perpetrators and victims remain in the relation of formal power or informal domination:

> 'So there have been situations where I wish I had stood up for myself more, but it is difficult when it is a person in a higher position, when you know there is some kind of ... "I am organizing a meeting dinner, they'll go, if you'll be going, I'll join". It's not really harassment, but it is uncomfortable. ... And I wish I could say something against it, but you know it's that moment when you wish you'd have said something back, but you have to be careful. These are the people giving you jobs.' (10_F)

Perceiving women as sexual objects means at the same time not treating them seriously as scientists, which our respondents are also aware of: "Either they think 'Oh, what does she want from me?' or they think 'Yes, what does that bimbo there want. She is not able to do anything anyways' or they just look at how you are dressed" (07_F).

Other themes of microaggressions

In the original conceptualization of gender microaggressions, three other themes were identified. These were sexist language, denial of the reality of sexism, and denial of individual sexism (Sue, 2010). Sexist language refers to using terms that infer superiority of men; for example, as in the phrase 'lady physics' (Barthelemy et al, 2016). However, in our material this problem did not come out as an independent factor; rather, women occasionally cited sexist terms when they discussed other cases of microaggression, including presumed incompetence or sexual subjectification (such as "neither fish nor fowl" depicting an attitude towards interdisciplinary research a woman physicist was involved in; "sweet blonde" or "bimbo" referring to descriptions of women physicists).

Denial of the reality of sexism means not believing that sexism exists and refers to situations when individuals are persuaded on different occasions that gender inequality is not an important problem, its scale is minimal, or is being sometimes instrumentalized to conceal professional failures of women (Barthelemy et al, 2016). Similarly, denial of individual sexism refers to situations when individuals deny personal sexist beliefs or behaviours. Indeed, a few interviewees signal being faced with a microinvalidation manifesting itself in lack of reaction from their superiors when cases of gender discrimination or harassment were disclosed as well as in lenient treatment of the perpetrators.

Perceived effects of microaggressions

When sharing experiences of unequal treatment at work, interviewees often also formulated their personal reflections about the effects of microaggressions on their lives. Interviewed women sometimes deny the negative impact of microaggressions due to their incidental character and own resilience to sexist comments. They also share their doubts whether what they had experienced was because of their gender and whether they are maybe 'too sensitive' to pay attention to the cases of misconduct, such as sexual objectification. However, when repetitive and ubiquitous, regardless of whether this occurs in the context of power relations or is performed by co-workers, being challenged by microaggressions is described as frustrating and exhausting. Being their target, women report feeling that they must work harder to prove to others that they are professional. The cumulation of sexism towards one of the interviewees even makes her have doubts as to whether she wants to continue her career in science:

'for 10 years I am working so much, so that I can do what I like, but at a certain point you are done with that. So, it is sort of *on the verge* right now. Because eventually you realize that the entire world is against you somehow. And that is exhausting. And you, you can, so you can convince yourself that it's all just small things and that this is not necessarily because you are a woman and blah, but eventually it will get too exhausting.' (07_F)

While the nature of microaggressions is ambiguous and various experiences of sexism are kept silent until they manifest in overt discrimination, such as mobbing or sexual harassment, women physicists admit developing various coping strategies. They include building up personal resilience as in the case of a senior woman physicist who reports "having learned not to worry about opinions" (74_F). Another strategy that is mentioned in the interview narratives is sharing experiences with trusted co-workers or family members, who offer comfort and advice. Finally, the interviewees openly describe taking measures aiming at assimilating to the masculine culture of physics. One of them recalls her graduate studies and membership in a physics students' organization, when she managed to get full acceptance in a male team and became one of the boys at the expense of her femininity:

'I had this feeling ... that if I wanted to be perceived as clever I had to be perceived as a man and that I had to become a man ... thanks to that I know jokes that I will never tell anyone in my life, I know what it's like in the army, here I had a crisis when it comes to being a woman, I was treated like a man, shoulder to shoulder and on the one hand I knew I had to give up something, and on the other hand it was such a unique time.' (55_F)

A similar technique is currently used by another respondent, who reports making deliberate efforts to resemble men through dress in order to be treated as a professional: "I'm determined to wear trousers, like the guys do, and it might seem pointless, but it's just so that they don't treat me differently, that I have a short skirt, so that there are even any subliminal or subconscious signals" (56_F).

Coping strategies undertaken by some of the interviewed women may as well include, as was suggested earlier, denying the very existence of sexism in physics, which has been claimed with conviction by a dozen of interviewed physicists.

Conclusions

The study confirms that European women physicists face particular microaggressions at different career stages, during their graduate education, as early career scientists, and when in more advanced academic positions. The most prevalent themes that have been identified in the analysed narratives include presumed incompetence, restrictive gender roles, invisibility, sexist jokes, and sexual objectification. Women's experiences vary from identifying individual incidents taking place in the past to having constantly been challenged with various variants of microaggressions (and overt discrimination, which is not discussed in this chapter). While some of the interviewees doubt that those microaggressive acts have any impact on their well-being and/or careers, other share negative consequences of being exposed to microaggressions. These include feeling bad and experiencing frustration about being treated differently than a person would like to be treated, feeling obliged to constantly provide evidence of being equally competent as men, and questioning one's ability to continue with an academic career. In order to cope with instances of covert sexism, the interviewed women employ various strategies, which – while bringing some individual advantages – hardly allow them to challenge the very culture of physics which, under the guise of 'neutrality', is overwhelmingly masculine.

As in other studies on gender microaggressions in sciences – for example, Barthelemy et al (2016) – there is some overlap in the analysed interviews between the discussed themes. For example, sexist jokes most often denote sexual objectification of women and sometimes also imply the assumption of their incompetence or inferiority. Similarly, restrictive gender norms and presumed incompetence often merge with each other as the belief that women should limit their activity to their traditional gender roles might be further strengthened by the conviction that they simply are unable to perform 'manly' tasks. While the overlapping of themes is a certain limitation in the use of the framework of microaggressions, at the same time it captures 'how small, often unconscious, behaviors can work to convey sexist messages to women', and that these are often multiple in form (Barthelemy et al, 2016: 10). Together with overt sexism, which was also reported in some of the analysed narratives, they create a hostile and invalidating climate for women physicists. Due to their subtle and veiled nature, identifying and countering microaggressions are difficult. While seemingly harmless, they can accumulate over time and hinder women's careers. Hence, learning to recognize various microaggressions or non-events would help women scientists to effectively respond to them. Potential tools for change could include anonymous pooling and publication of the experiences of microaggressions, monitoring the practices of support, encouragement, inclusion, and exclusion in research groups, projects, networks, conferences, and science institutions from a gender perspective, addressing the issue of microaggressions both in leadership and management training and early career coaching for both men and women (Husu, 2013). The existence of gender bias that is the basis for microaggressions and overt forms of gender discrimination is more likely in heavily male-dominated fields (ILO, 2017), which is true for physics. Raising awareness in the field about the role of implicit gender bias in the processes of recruitment, evaluation, and recognition of merit and about the effects of overt and covert forms of gender discrimination on women's careers becomes an important element of systemic efforts to transform physics into an inclusive environment that affirms both men's and women's belonging.

Notes

[1] Besides, physics is one of the fields whose practitioners believe that raw, innate talent is the main requirement for success and women are stereotyped as not possessing such talent (Leslie et al, 2015).

[2] GENERA was a H2020 project (2015–18) funded by the European Commission under GERI-4-2014 grant agreement 665637.

[3] In the whole sample there were 44 semi-standardized interviews with women physicists; however, in four cases an issue of unequal treatment and gender discrimination was not raised. Therefore, these interviews are not included in this analysis.

[4] In one partner institution (Spain) the study was conducted on the basis of a slightly different scenario designed by a local project team. Most of the topics raised in both scenarios were, however, analogous, which allowed for including data from Spain into the common analysis.

[5] A question "Do you think you were discriminated against or treated worse than men in some situations because you are a woman? Please, give me some examples" was directly asked if the respondents did not share any such incidents in their accounts of becoming

a physicist and pursuing a career in physics. If they did, the interviewers referred to these experiences and asked some specific follow-up questions to receive better understanding of their nature and consequences.

[6] Additionally, one interviewee admitted that she was not sure whether she had been discriminated or not and two interviewees gave very vague descriptions of their experiences, which precludes their categorization. It is also worth mentioning that a few respondents reported instances of overt discrimination understood as explicit, direct, intentional, and often-unlawful forms of negative demeanour and/or treatment towards the members of social minority on the basis of their minority status membership (Jones et al, 2016: 1591). These incidents included mobbing, sexual harassment, and discrimination in access to resources as well as unequal pay; however, their discussion goes beyond the scope of this chapter.

References

Antolini, R., Cenci, P., Croci, S., Leone, S., Masullo, M.R., Picardi, I., and Trinchieri, G. (2019) 'Women and physics in Italy: numbers, projects, actions', *AIP Conference Proceedings 2109*, 050023, DOI: 10.1063/1.5110097

Aycock, L.M., Hazari, Z., Brewe, E., Clancy, K.B., Hodapp, T., and Goertzen, R.M. (2019) 'Sexual harassment reported by undergraduate female physicists', *Physical Review Physics Education Research*, 15(1): 010121.

Barthelemy, R., McCormick, M., and Henderson, C. (2014) 'Understanding women's gendered experiences in physics and astronomy through microaggressions', *2014 Physics Education Research Conference Proceedings*, Minneapolis, MN.

Barthelemy, R., McCormick, M., and Henderson, C. (2016) 'Gender discrimination in physics and astronomy: graduate student experiences of sexism and gender microaggressions', *Physical Review Physics Education Research*, 12(2): 020119.

Benschop, M. and Brouns, M. (2003) 'Crumbling ivory towers: academic organisations and its gender effects', *Gender, Work and Organisation*, 10(2): 194–212.

Berk, R.A. (2017) 'Microaggressions trilogy: Part 1. Why do microaggressions matter?', *Journal of Faculty Development*, 31(1): 63.

Blithe, S.J. and Elliot, M. (2020) 'Gender inequality in the academy: microaggressions, work–life conflict, and academic rank', *Journal of Gender Studies*, 29(7): 751–64.

Carli, L.L., Alawa, L., Lee, Y., Zhao, B., and Kim, E. (2016) 'Stereotypes about gender and science: women scientists', *Psychology of Women Quarterly*, 40(2): 244–60.

Ceci, S.J. and Williams, W.M. (2010) 'Understanding current causes of women's underrepresentation in science', *Proceedings of the National Academy of Sciences*, 108(8): 3157–62, Available from: www.pnas.org/content/108/8/3157.full.pdf

Ceci, S.J., Ginther, D.K., Kahn, S., and Williams, W.M. (2014) 'Women in academic science: a changing landscape', *Psychological Science in the Public Interest*, 15(3): 75–141.

Charles, M. and Bradley, K. (2009) 'Indulging our gendered selves? Sex segregation by field of study in 44 countries', *American Journal of Sociology*, 114(4): 924–76.

Cheryan, S., Ziegler, S.A., Montoya, A.K., and Jiang, L. (2017) 'Why are some STEM fields more gender balanced than others?', *Psychological Bulletin*, 143(1): 1.

Cronin, C. and Roger, A. (1999) 'Theorizing progress: women in science, engineering, and technology in higher education', *Journal of Research in Science Teaching*, 36(6): 637–61.

De Hoogh, A., Hesping, S., Rudolf, P., and de Wolf, E. (2019) 'The Dutch Fom/f approach to gender balance in physics', *AIP Conference Proceedings 2109*, 050028. DOI: 10.1063/1.5110102

De Welde, K. and Laursen S. (2011) 'The glass obstacle course: informal and formal barriers for women PhD students in STEM fields', *International Journal of Gender, Science and Technology*, 3(3): 572–95.

Elsevier (2017) *Gender in the Global Research Landscape*, Available from: www.elsevier.com/__data/assets/pdf_file/0003/1083945/Elsevier-gender-report-2017.pdf

Eran Jona, M. and Nir, Y. (2021) 'Choosing physics within a gendered power structure: the academic career in physics as a "deal"', *Physical Review Physics Education Research*, 17(2): 020101.

Estacio, E.V. and Saidy-Khan, S. (2014) 'Experiences of racial microaggression among migrant nurses in the United Kingdom', *Global Qualitative Nursing Research*, 1: 2333393614532618.

Etzioni, A. (2014) 'Don't Sweat the Microaggressions. The old pitfalls of new sensitivities in political speech', *Atlantic*, 8 April.

European Commission (2021) *She Figures 2021*, Luxembourg: Publications Office of the European Union.

Flick, U. (2006) *An Introduction to Qualitative Research*, London; Thousand Oaks, CA: Sage.

Fotaki, M. (2013) 'No woman is like a man (in academia): the masculine symbolic order and the unwanted female body', *Organization Studies*, 34(9): 1251–75.

Friedlaender, C. (2018) 'On microaggressions: cumulative harm and individual responsibility', *Hypatia*, 33(1): 5–21.

Gaisch, M., Chydenius, T., Preymann, S., Sterrer, S., and Aichinger, R. (2016) 'Gender microaggressions in low – context communication cultures: a perceptual study in the context of higher education institutions', *Proceedings Cross-Cultural Business Conference*, 19–20 May, Steyr, Austria.

Gartner, R.E. (2021) 'A new gender microaggressions taxonomy for undergraduate women on college campuses: a qualitative examination', *Violence Against Women*, 27(14): 2768–90.

Gonsalves, A.J., Danielsson, A., and Pettersson, H. (2016) 'Masculinities and experimental practices in physics: the view from three case studies', *Physical Review Physics Education Research*, 12(2): 020120.

Guillopé, C. and Roy, M.F. (2020) *A Global Approach to the Gender Gap in Mathematical, Computing, and Natural Sciences. How to Measure It, How to Reduce It?* Gender Gap in Science project, final report. DOI: 10.5281/zenodo.3882609

Haas, M., Koeszegi, S.T., and Zedlacher, E. (2016) 'Breaking patterns? How female scientists negotiate their token role in their life stories', *Gender, Work and Organization*, 23(4): 397–413.

Harris, M. (2016) 'A thousand tiny cuts', *Physics World*, 29(3): 45.

Holman, L., Stuart-Fox, D., and Hauser, C.E. (2018) 'The gender gap in science: how long until women are equally represented?', *PLoS Bio*, 16(4): 1–20.

Hughes, R. (2014) 'The evolution of the chilly climate for women in science', in B. Irby, B. Polnick, and J. Koch (eds) *Girls and Women in STEM: A Never-Ending Story*, College Park, MD: Information Age Publishing, pp 71–92.

Husu, L. (2013) 'Recognize hidden roadblocks', in L. Al-Gazali, V. Valian, B. Barres, L. Wu, E. Andrei, J. Handelsman, C. Moss-Racusin, and L. Husu (eds) 'Scientists of the world speak up for equality', *Nature*, 495(7439): 35–8.

Husu, L. (2020) 'What does not happen: interrogating a tool for building a gender-sensitive university', in E. Drew and S. Canavan (eds) *The Gender-Sensitive University: A Contradiction in Terms?*, London and New York: Routledge, pp 166–76.

ILO (2017) *Breaking Barriers: Unconscious Gender Bias in the Workplace*, Available from: www.ilo.org/wcmsp5/groups/public/---ed_dialogue/---act_emp/documents/publication/wcms_601276.pdf

Ivie, R. and Ray, K.N. (2005) *Women in Physics and Astronomy*, College Park, MD: American Institute of Physics.

Jones, K.P., Peddie, C.I., Gilrane, V.L., King, E.B., and Gray, A.L. (2016) 'Not so subtle', *Journal of Management*, 42(6): 1588–613.

Keller, E.F. (2001) 'The anomaly of a woman in physics', in M. Wyer, M. Barbercheck, D. Geisman, H.O. Ozturk, and M. Wayne (eds) *Women, Science, and Technology: A Reader in Feminist Science Studies*, New York: Routledge, pp 9–16.

Krzaklewska, E., Sekuła, P., Ciaputa, E., and Struzik, J. (2019) 'Why does it happen in physics? Opinions of European physicists on gender inequality', *Studia Humanistyczne AGH*, 18(4): 13–30.

Leslie, S.J., Cimpian, A., Meyer, M., and Freeland, E. (2015) 'Expectations of brilliance underlie gender distributions across academic disciplines', *Science*, 347(6219): 262–5.

Lewis, K.L., Stout, J.G., Finkelstein, N.D., Pollock, S.J., Miyake, A., Cohen, G.L., and Ito, T.A. (2017) 'Fitting in to move forward: belonging, gender, and persistence in the physical sciences, technology, engineering, and mathematics (pSTEM)', *Psychology of Women Quarterly*, 41(4): 420–36.

McClure, E. and Rini, R. (2020) 'Microaggression: conceptual and scientific issues', *Philosophy Compass*, 15(4): e12659.

McCullough, L. (2019) 'Women in physics leadership', *AIP Conference Proceedings 2109*, 130006. DOI: 10.1063/1.5110154

McCullough, L. (2020) 'Barriers and assistance for female leaders in academic STEM in the US', *Education Sciences*, 10(10): 264.

Miikkulainen, K., Ott, J., and Vapaavuori, J. (2019) 'Update on women in physics in Finland', *AIP Conference Proceedings 2109*, 050015. DOI: 10.1063/1.5110089

Miner, K.N., January, S.C., Dray, K.K., and Carter-Sowell, A.R. (2019) 'Is it always this cold? Chilly interpersonal climates as a barrier to the well-being of early-career women faculty in STEM', *Equality, Diversity and Inclusion: An International Journal*, 38(2): 226–45.

Napier, J.L., Suppes, A., and Bettinsoli, M.L. (2020) 'Denial of gender discrimination is associated with better subjective well-being among women: a system justification account', *European Journal of Social Psychology*, 50(6): 1191–209.

O'Connor, P. (2020) 'Why is it so difficult to reduce gender inequality in male-dominated higher educational organizations? A feminist institutional perspective', *Interdisciplinary Science Reviews*, 45(2): 207–28.

O'Connor, P., O'Hagan, C., and Gray, B. (2018) 'Femininities in STEM – outsiders within', *Work, Employment and Society*, 32(2): 312–29.

O'Connor, P. Carvalho, T., Vabø, A., and Cardoso, S. (2015) 'Gender in higher education: a critical review', in J. Huisman, H. de Boer, D.D. Dill, and M. Souto-Otero (eds) *The Palgrave International Handbook of Higher Education Policy and Governance*, Basingstoke: Palgrave, pp 569–85.

Ong, M., Smith, J.M., and Ko, L.T. (2018) 'Counterspaces for women of color in STEM higher education: marginal and central spaces for persistence and success', *Journal of Research in Science Teaching*, 55(2): 206–45.

Periyakoil, V.S., Chaudron, L., Hill, E.V., Pellegrini, V., Neri, E., and Kraemer, H.C. (2019) 'Common types of gender-based microaggressions in medicine', *Academic Medicine*, 95(3): 450–7.

Piccinelli, E., Martinho, S., and Vauclair, C.M. (2020) 'What kinds of microaggressions do women experience in the health care setting? Examining typologies, context and intersectional identities', Available from: https://repositorio.iscte-iul.pt/bitstream/10071/20672/1/6706-Article%20Text-25929-1-10-20200713.pdf

Porter, A.M. and Ivie, R. (2019) *Women in Physics and Astronomy*, the Statistical Research Center of the American Institute of Physics.

Rolin, K. (2008) 'Gender and physics: feminist philosophy and science education', *Science and Education*, 17(10): 1111–25.

Rolin, K. and Vainio, J. (2011) 'Gender in academia in Finland: tensions between policies and gendering processes in physics departments', *Science Studies*, 24(1): 26–46.

Rossiter, M.W. (1993) 'The Matthew Matilda effect in science', *Social Studies of Science*, 23(2): 325–41.

Šatkovskiene, D., Kupliauskiene, A., Rutkuniene, Ž., and Ružele, Ž. (2019) 'Gender equality in research: Lithuania is starting modernization of research organizations performing physics research', *AIP Conference Proceedings 2109*, 050025. DOI: 10.1063/1.5110099

Savigny, H. (2019) 'Cultural sexism and the UK Higher Education sector', *Journal of Gender Studies*, 28(6): 661–73.

Sekuła, P., Struzik, J., Krzaklewska, E., and Ciaputa, E. (2018) *Gender Dimensions of Physics: A Qualitative Study from the European Research Area*, Available from: https://genera-project.com/portia_web/Gender_Dimensions_of_Physics.pdf

Simatele, M. (2018) 'A cross-cultural experience of microaggression in academia: a personal reflection', *Education as Change*, 22(3): 1–23.

Stojanović, M., Pavkov-Hrvojević, M., Božić, M., Knežević, D., Davidović, M., Trklja, N., Žekić, A., Marković Topalović, T., and Jovanović, T. (2019) 'Gender imbalance in the number of PhD physicists and in key decision-making positions in the Republic of Serbia', *AIP Conference Proceedings 2109*, 050033. DOI: 10.1063/1.5110107

Sue, D.W. (2010) *Microaggressions in Everyday Life: Race, Gender and Sexual Orientation*, Hoboken, NJ: John Wiley & Sons.

Sue, D.W. and Spanierman, L. (2020) *Microaggressions in Everyday Life*, Hoboken, NJ: John Wiley & Sons.

Timmers, T.M., Willemsen, T.M., and Tijdens, K.G. (2010) 'Gender diversity policies in universities: a multi-perspective framework of policy measures', *Higher Education*, 59: 719–35.

Traweek, S. (1988) *Beamtimes and Lifetimes: The World of High Energy Physicists*, Cambridge, MA; London: Harvard University Press.

Wennerås, C. and Wold, A. (1997) 'Nepotism and sexism in peer-review', *Nature*, 387(6631): 341–3.

Yang, Y. and Carroll, D.W. (2018) 'Gendered microaggressions in science, technology, engineering, and mathematics', *Leadership and Research in Education*, 4: 28–45.

Zippel, K.S. (2017) *Women in Global Science: Advancing Academic Careers through International Collaboration*, Stanford, CA: Stanford University Press.

6

The Physics PhD Race and the 'Glass Hurdles' for Women: A Case Study of Israel

Meytal Eran Jona and Yosef Nir

Introduction

This chapter builds on the research presented in Chapter 4 which examined the challenges faced by women physics researchers in Israel at the postdoctoral career stage. In this chapter, we turn to look at the doctoral stage itself to investigate the gender issues which may be experienced by Israeli women seeking academic careers in physics. As we discussed in Chapter 4, the proportion of women who acquire higher education in the West has been increasing steadily from the 1950s, such that nowadays, they constitute a majority among undergraduate and graduate students in many disciplines. In light of this fact, the lack of women among students and academic staff in science, technology, engineering, and mathematics (STEM) has been the focal point of research and action in Western democracies for the last two decades (Sarseke, 2018).

The contemporary picture in Israel is particularly poor. Over the last decade, women constitute 18 per cent of all physics BSc, MSc, and PhD students, and at present they represent only 6 per cent of the academic staff within Israeli universities (Eran Jona and Nir, 2019). In contrast, in life sciences, the picture is different. For instance, in medicine, women constitute 69 per cent of all PhD graduates and 35 per cent of the academic staff, and in biology, women constitute 58 per cent of all PhD graduates and 30 per cent of the academic staff. Women's success in these sciences raises important questions for the disciplines in which they are so poorly represented.

Doctoral students are an under-studied group that is of particular interest in the context of investigating the gender gap in STEM fields and academic careers. Previous research reveals a high attrition rate for women before and during postdoctoral studies, a key period towards academic careers, where the numbers of women decrease dramatically (Carmi, 2011; Gofen, 2011; Goulden et al, 2011; Bostwick and Weinberg, 2018). With the gender gap in physics so severe in Israel, our research questions were: How do women experience doctoral studies? Do they face different difficulties than men? We used a representative nationwide survey among all PhD students in Israel to answer these questions.

Israel provides an interesting context in which to undertake the study for several reasons. First, Israel is a small country, particularly compared with the US and the UK, where the existing knowledge is primarily based. Thus, we can reach almost the entire population of the physics community, rather than sample it. Second, the Israeli physics community is an inseparable part of the vibrant international community, following the cultural norms that construct the field. On the other hand, Israel has some unique characteristics: Israelis on average marry at a relatively young age and have more children than in other Western societies; most Israeli women maintain a full-time employment history, yet the division of labour at home is still unequal. These similarities and differences make Israel an interesting 'laboratory' for gender in physics. For more details, see Chapter 3.

Our findings reveal both the shared as well as the different experience of women along the PhD track. They also highlight how Israeli women experience their physics PhD studies as a series of hurdles, albeit these are 'glass hurdles' as they are hidden to the academic system. From the findings, explanation can also be given for the minority of women in the field, based on the additional obstacles they have to face. In the conclusions, we will also suggest courses of action to improve the situation.

Background

In the following, we build on the literature reviewed in Chapter 4 to explore what is known regarding obstacles and difficulties that physics graduate students specifically face alongside their studies, with some references to other STEM fields and to research on academic careers more generally.

For a more detailed discussion of the theoretical framework within which our analysis and explanations are embedded, see Chapter 4.

The culture of physics as a male-dominated field

As we discussed in Chapter 4, Traweek (1988) was one of the first social scientists that focused her research on the physics community. She studied the culture of physics in high-energy laboratories in the US and Japan, and found communities based on intense competition. The available resources are limited, and their distribution is often based on social connections and biased decisions. This is in stark contrast to the imagined culture of the discipline, which many see as the pinnacle of rationality, empty of emotion, and devoid of human influence.

The culture of physics as a competitive and male-dominated field that Traweek observed may imperil women's participation in the field. If women are seen as contrary to science – particularly the 'fundamental' and 'objective' science of physics – then they may be immediately seen by the gatekeepers of science and community members as being 'unfit'. As stated previously, physics is associated with rationality, objectivity, and logic: features that have been historically associated with masculinity (Harding, 1991).

Three decades after Traweek's work, research into diverse physics sub-fields in various countries reveals that the culture of physics as a male-dominated field in Western democracies persists. The masculine work culture, alongside growing demands for competitiveness and career dedication, pose obstacles for women. Moreover, reconciling work and private life becomes more difficult in a more precarious model of a career demanding mobility, and brings new challenges for partners in dual-career couples (Sekula et al, 2018). Women in the field of physics have not only to navigate the masculine norms of the discipline, but also to negotiate the limited possible identities for women in physics (Gonsalves et al, 2016).

Discussing power relations within the academy as a whole, Bagilhole and Goode (2001) claim that there is a fixed, standard academic career model that is not gender-neutral, but is based on a masculine model anchored in hegemonic masculine culture and a patriarchal support system. In a study of MIT senior women faculty, it was found that the 'ideal' perfect academic is one who gives total priority to career and has no outside interests and responsibilities (Bailyn, 2003). What we see in physics is a vicious circle, whereas the absence of women physicists generates reluctance among women to make the effort to fit into a field where they will be a marginal minority. Physics as a masculine field is one of the persisting 'castles' of gender imbalance.

The perception of a profession as male or female is also influenced by the extent to which an occupation allows or does not allow combining family life with a career. In a study of women who completed their PhD in STEM fields with excellent grades, this component was found to play a significant role in the decision-making of whether to pursue an academic career in science (Gofen, 2011).

As discussed in Chapter 4, Lamont and Molnár (2002) explain the preservation of segregation between women and men via the term 'boundaries'. Boundaries are complex structures – physical, social, and psychological – that produce commonalities and differences between women and men, and in turn shape and structure the behaviour and attitudes of each gender (Gerson and Peiss, 1985). Social boundaries are used to distinguish between women and men in the workplace. Thus, male employees are perceived as more competent than women employees are. Those who violate gender boundaries and accepted norms, such as the norm of dedicating yourself and all your time to work, may suffer from stigma and punishment in the workplace (Epstein, 2004). Looking at physics departments worldwide, it seems that there are clear boundaries that prevent many talented women from choosing a career in physics, and that those boundaries are closely related to the culture of physics as a masculine field.

With our interest in the doctoral stage, in what follows we review the literature on the difficulties that physics (and, more generally, maths-intensive) PhD students may encounter during their studies, with emphasis on gender aspects.

Mental health within faculty and PhD students

Research on undergraduate and graduate students in all disciplines shows higher rates of mental health issues among students compared with the overall population. Literature reviews of mental health in research environments (Guthrie et al, 2018) indicate that PhD students in particular face mental health issues. The proportions of both university staff and postgraduate students with a risk of having or developing a mental health problem, based on self-reported evidence, were generally higher than for other working populations. Moreover, large proportions (more than 40 per cent) of postgraduate students in the UK report symptoms of depression, emotion- or stress-related problems, or high levels of stress. The main factors associated with the development of depression and other common mental health problems in PhD students are high levels of work demands; the pressure to publish and win grants in highly competitive environments; job insecurity; work–life conflict; little say in or control over their work; poor support from their supervisor; and exclusion from decision-making. Believing that PhD work is valuable for one's future career helps reduce stress, as does confidence in one's own research abilities.

Furthermore, gender was found to be the key personal factor that contributes to mental health outcomes in the workplace. Women report more exposure to stress than men do. Moreover, they also report greater challenges around work–life balance. The results on whether age affects mental health were inconclusive, partly because age is often difficult to disentangle from the discussions about rank and seniority. Other factors such as disability, sexuality, and minority status were mentioned in a small number of articles, which indicate that these personal factors generally increase stress (Guthrie et al, 2018).

While this research refers to the general academic environment, it is plausible that the findings are particularly relevant to the very competitive field of physics. Indeed, research conducted among PhD and MSc students within one of the leading Israeli academic institutes reveals that a higher percentage (34 per cent) of physics students report that their mental health was negatively affected by their studies compared with students in other science faculties (26–29 per cent) (Eran Jona and Perez, 2021).

Work–life conflict in the academy

Work–life conflict is a source of stress related to workload. A survey conducted among all active members of the UK's University and College Union (UCU) and reported by Kinman (2013) shows that work demands are the strongest predictor of work–life conflict. In that survey, the majority of respondents reported that their ideal level of work–life separation would be greater than what they experienced at the time of reporting. Tytherleigh et al (2005) also found that work–life conflict and work overload were sources of stress for higher education staff, but that the stress levels associated with these stressors were lower than for individuals working in other areas (Guthrie et al, 2018). According to a study of women academics in a UK university, the ultimate responsibility for children and the elderly still rests on women's shoulders. Close to half believe that family responsibilities have interfered with their career progression. The findings indicate that in departments dominated by men, there was less empathy with the non-work responsibilities of women (Forster, 2001).

In a Belgian study, work–life conflict was identified among PhD students as the most important predictor of mental health problems, followed by work demands (Levecque et al, 2017). This factor was also identified as important in a UK study of PhD students, which found that having a high workload that affects private life was a bothersome issue for respondents (Hargreaves et al, 2014).

Research focused on a worldwide sample of physicists indicates that, by an almost 2:1 margin, women are more likely than men to say that becoming a parent significantly affected their work in various ways. Women were most likely to report changing their schedules, spending less time at work, and becoming more efficient (when having children). Those findings echo results from the first two International Union of Pure and Applied Physics (IUPAP) surveys (conducted in 2002 and 2005), in which women physicists reported that having children forced them to become more efficient because they had to leave their laboratory or office in time to pick up young children from childcare. The respondents were also asked whether their employers assigned less challenging work to them when they became parents. The majority of physicists did not report a change. Still, women were more likely than men to report being given less challenging work, and the difference was statistically significant (Ivie and Tesfaye, 2012). Nevertheless, women are not opting out of the workforce for family reasons alone (Cabrera, 2007).

Gender-based discrimination in the academy

Sexism occurs when men are believed to be superior to women. It is considered one of the reasons for women's under-representation in physics. Sexism was studied in many fields, including higher education, and was found to be positively correlated with anxiety symptoms, stress, and role overload (West, 2014). The issue of sexism in physics and astronomy has not been thoroughly explored in the literature and there is currently neither much evidence for it, nor even clear language by which to discuss it. Barthelemy et al (2016) led one of the few relevant research projects. It was based on interviews with women in graduate physics and astronomy programmes, and explores their individual experiences of sexism. Although a minority of the women interviewed did not report experiencing sexual discrimination, the majority experienced subtle insults and microaggressions. Other participants also experienced more traditional hostile sexism in the form of sexual harassment, gender role stereotypes, and overt discouragement.

'Microaggressions' is a term describing a subtle form of gender bias. Among the dominating themes or forms, one finds sexual objectification, second-class citizenship, use of sexist language, assumption of inferiority, restrictive gender roles, invisibility, and sexist jokes, as well as denial of the reality of sexism (Sue, 2010; Barthelemy et al, 2016). It is argued that microaggressions 'act upon women in several ways, by reiterating the social view that men are more valued than women, by reinforcing traditional stereotypes about proper gender roles, and by contributing to violence toward women by objectifying and sexualizing them' (Barthelemy et al, 2016: 4). Therefore, the consequences of microaggressions may be as severe as those of overt sexism.

Research has found that women physicists, including graduate students and faculty, frequently encounter microaggressions. Interviews with physicists (44 female and 22 male) from 12 research institutions in eight European countries indicate that women in academic careers in physics face various forms of microaggressions. Furthermore, women physicists more often declare being unequally treated in their workplace than their male counterparts do. The significance of microaggressions is that it signals deprecation of the professional position of women physicists. Microaggressions evoke negative emotions in women and their accumulation may contribute to women leaving science (Sekula et al, 2018). This important issue is discussed more fully by Sekula in Chapter 5.

Sexual harassment in the academy

Sexual harassment is a form of gender discrimination. Broadly defined, sexual harassment is unwelcomed or inappropriate behaviour of a sexual nature that creates an uncomfortable or hostile environment. It comes in various forms, both subtle and overt. A study of sexual harassment in physics (Aycock et al, 2019) considers three specific types. First, 'sexist gender harassment' describes hostile or insulting remarks and actions based on one's gender, such as saying that women cannot do physics. Second, 'sexual gender harassment' refers to sexual remarks or conduct, such as commenting on the shape of a woman's body. Third, 'unwanted sexual attention', such as requests for sexual favours or unwanted touching.

The data regarding the extent of sexual harassment in the academy vary. Nevertheless, following the corpus of research in the field, sexual harassment appears to be an epidemic throughout global higher education systems, which impacts on individuals, groups, and entire organizations in profound ways. Bondestam and Lundqvist (2020), summing up the core results from the most highly cited research papers in scientific journals, conclude that exposure to sexual harassment in higher education varies between 11 and 73 per cent for heterosexual women (median 49 per cent) and between 3 and 26 per cent for heterosexual men (median 15 per cent).

Research has shown that women in male-dominated occupations are at greater risk of being sexually harassed, and that these experiences increase job turnover intentions and withdrawal from work. Studies focused on women in science, technology, engineering, mathematics, and medicine (STEMM) fields indicate that sexual harassment affects the majority of women (Clancy et al, 2014; Lorentz et al, 2019). There is little research evidence on sexual harassment in physics specifically. A survey of undergraduate women physics students has shown that approximately three quarters (74 per cent) of survey respondents experienced at least one type of sexual harassment (Aycock et al, 2019). It was also found that certain types of sexual harassment in physics predict a negative sense of belonging and exacerbate the imposter phenomenon (Aycock et al, 2019).

These findings are not surprising, given previous studies showing that experiencing sexual harassment (in general) increases a woman's likelihood of leaving a STEM career (Johnson et al, 2018). For those women who do stick with their field, harassment hurts their career, economic standing, and well-being (Howe-Walsh and Turnbull, 2016). In short, unchecked harassment creates a drain of talent through lost work, lost ideas, and lost people (Libarkin, 2019).

Against this background, our research aimed to explore gender differences within physics PhD students in their attitudes regarding academic studies and their personal experience. We now turn to discuss this in more detail.

Research study

The research methods we selected for the study are embedded within feminist research approaches and theories that provide frameworks and tools for looking into women's lives (Reinharz and Davidman, 1992; DeVault, 1999; Krumer-Nevo et al, 2014). In order to learn physicist women's points of view, we conducted in-depth interviews with women PhD students, postdoctoral fellows, and academic staff members. Based on insights from the interviews, we constructed a questionnaire, which was then distributed to all (female and male) physics PhD students in Israel in order to widen our understanding of the gender-related differences. The interviews provide the basis for our report in Chapter 4. Here we focus on the findings from the nationwide survey of physics PhD students in Israel.

Our research team is unique as it consists of a woman sociologist and a man physicist. The sociologist has had previous experience in studies of gender, with a career spanning two decades. As an organizational sociologist she studied various aspects of gender and family in male-dominated organizations and served as a gender-equality advisor and diversity and inclusion officer in an institution of higher education. She gained relevant experience serving as the chairwoman of a European project to promote gender equality in physics (GENERA network, see Chapter 7). The physicist, beyond his experience in theoretical particle physics research, gained some perspective of the cultural norms and the power structure of the field, by serving as dean of a physics faculty, a chairperson of an institutional promotion committee, and a member in various international advisory committees. Together we decided to step out of our respective disciplinary comfort zones and collaborate in a long-term research project on gender aspects in physics.

We aimed to gain a deeper understanding of the reasons for the under-representation of women in physics to be able to formulate practical recommendations to improve the gender balance in the field, and eventually have a real impact on the Israeli academy.

In terms of methods, the research team compiled an online self-administrated questionnaire that was sent directly via e-mail to university physics PhD students. The survey questionnaire was compiled by the research team in consultation with researchers at the American Institute of Physics (AIP), which has been researching student attitudes towards physics for a decade. The research questionnaire was partly based on the tools developed at the AIP for research in the field, while adapting it to the Israeli context and to the research questions that interested us. The questionnaire included 107 questions, of which six were open-ended. It consisted of questions regarding the following topics: the students' socio-demographic background, academic study track, attitudes regarding the academic environment, success indicators, combining family and studies, future employment expectations and intentions, desire to have an academic career, considerations in favour of and against postdoctoral studies, and aspects of discrimination and sexual harassment during academic studies.

In terms of population and sampling, the deans of physics at all (six) research universities in Israel contacted all PhD students (N=404) at their institutions to answer the questionnaire. Therefore, we addressed the entire population of physics students in Israel and not a sample of them. The research team made efforts to encourage all students to respond, and a few reminders were sent. Respondents (n=267) accounted for 66 per cent of the overall student population, of which 60 were women and 207 men. The research team made a special effort to raise women's response rates because of the small population and the researchers' interest in this group. These efforts resulted in a 94 per cent women response rate (N=64, n=60).

The maximum margin of error for the entire population is 3.6 per cent, for women 3.2 per cent, and for men 4.3 per cent. Due to the over-representation of women in the sample, the total number of students was weighted by gender; the data for the entire sample included in the study are weighted.

Results

Common difficulties regardless of gender

One of the most interesting findings is the similarities we found between the young men and women in describing the academic path and its challenges.

One of the questions in the survey aimed to find out what are the difficulties and struggles which all students, regardless of their gender, face during their studies. To achieve this goal, we asked an open-ended broad question: "If a close friend were to consult with you about PhD studies in physics, which difficulties would you present to them?" The indirect formulation of this question was aimed to examine what, in the students' eyes, are the main difficulties, independent of whether they experienced them in person. Most students had an answer to this question, many with a detailed explanation (221/267). Based on qualitative analysis, we found three main areas of difficulties: professional, economic, and personal.

Professional difficulties

The most common type of difficulties mentioned is difficulties related to the doctoral path, which is depicted as a busy, difficult, and frustrating course of study, which involves significant dependence on the supervisor and has elements of uncertainty. Study load was noted as one of the most prominent difficulties by many of the respondents (54), while the word "difficult" (21) and the word "frustrating" (10) were used repeatedly by the respondents in reference to the course of study.

Many mentioned the need to invest a lot of time, the fact that the track is very long (14), and talked about frustration from dealing with failures and lack of success which are an almost built-in part of the research process (15). The issue of time (43), as a scarce resource, appeared many times, while it was noted that the degree requires a great investment of time and time-management ability. Another difficulty was a competitive learning environment (10). Dependence on the supervisor (and not on the individual) as a condition for success was also mentioned as a difficulty (34).

Economic difficulties

The second most mentioned difficulty was economic. A considerable number of the respondents (44) mentioned the issue of low, unsafe, and insufficient fellowship as one of the main difficulties in the PhD. In addition, the lack of social conditions accompanying the scholarship, especially saving for retirement, and the lack of job security were noted negatively. Students said that the scholarship is not enough to cover their basic living expenses, that they ran into financial difficulties during their studies, and even described the study period as a period of financial "sacrifice". A large number of times (20) it was stated that low salary does not make it possible to support the family (spouse and children). Another difficulty noted in this context is an unclear employment horizon, and in particular the low chance of getting a tenure track position in academia (13).

Personal difficulties

The third area of difficulty noted as characterizing the doctoral period is the personal and emotional area. In this, loneliness was mentioned as one of the difficulties associated with the doctoral course. The requirement for independent work is an inseparable component of the PhD experience in many fields, as this is the student's first independent research work, and as such, it has an understandable degree of loneliness and responsibility that the student has not yet experienced during their studies. Therefore, it was not surprising to find that loneliness (14) and independence (13) were mentioned as a difficulty in describing the doctoral course. Independence was noted as a difficulty, because it includes both uncertainty and loneliness. In addition, some stated that it is difficult to conduct oneself without supervision independently (12), that time management is difficult, and that a lot of self-discipline and patience are required (11). Another major issue noted as a difficulty is uncertainty (25), both regarding research and the occupational future. Moreover, a number of respondents (14) stated that doctoral studies also involve an emotional burden, which causes mental stress. In this context, students reported anxiety and depression that they experienced themselves or that their friends experienced during the study and as a result of dealing with the demanding requirements of the PhD track.

Therefore, the first interesting finding was that in many aspects of the physics academic studies, women and men have the same experience. The PhD track in physics is perceived as difficult and highly demanding; the fellowship does not cover the students' needs at that stage of life. During the studies, both genders are experiencing loneliness and have to deal with uncertainties regarding their future career prospects.

Unique difficulties and challenges for women

While in many subjects the experience of students was similar, regardless of gender, it was interesting to find the areas where there were statistically significant differences between the answers of women and men in our study.

The first subject in which we found clear differences between women and men is in another series of closed questions that quantitatively tested the difficulties during the doctorate. In the survey we asked: "During your PhD studies, did you experience a period when it was difficult for you to provide what was professionally expected of you?" Of the women, 71 per cent answered positively, compared with 63 per cent of men ($p<0.05$). When asked a closed question about the reasons for this difficulty (see Table 6.1 for details), a much larger percentage of women (39 per cent) compared with men (23 per cent) mentioned health-related problems, both physiological and psychological.

In order to understand the significance of these data, one should refer to the uniqueness of the Israeli context, which has certain special characteristics. Israeli society is a familial society. Israelis marry on average at a relatively young age and have more children than in other Western democracies. Furthermore, because of the compulsory military service (2–3 years for both women and men starting after high school at the age of 18), Israeli

Table 6.1: Main reasons that prevented respondents from meeting PhD goals

	Women, *n=43*		Men, *n=130*	
	%	CI	%	CI
Pregnancy, raising children	42	[38,46]	40	[35,45]
Mental health	**34**	**[30,38]**	17	[13,21]
Physical health	**22**	**[19,25]**	9	[6,12]
Caring for family member	12	[9,15]	17	[13,21]
Crisis with spouse	10	[7,13]	20	[16,24]

Note: Data are presented by gender; CI stands for 95% confidence interval; bold indicates $p<0.05$.

students are older on average than their peers in other countries. Thus, in Israel, most of the PhD students already have a spouse. Among the survey respondents, 70 per cent were married or in relationships, and 40 per cent were already parents.

In a following question 95 per cent of mothers and 86 per cent of fathers stated that becoming parents affected their way of studying. A large majority of them (73 per cent) reduced the time spent on studies and research, and for a large fraction (34 per cent), this led to a reduced rate of progress in their research. Women mentioned more frequently than men that they learned to make their schedule more flexible (60 per cent versus 48 per cent, $p<0.05$) and to be more efficient and productive (40 per cent versus 27 per cent, $p<0.05$).

Because women are the ones to give birth, breastfeed, and take care of the newborn during parental leave (the Hebrew term translates, somewhat ironically, to 'birth vacation'), we assumed that combining pregnancy and parenthood with studies should be much more challenging to them. Indeed, when we asked about the parental leave, 69 per cent of the mother-students took a four-month leave (which is the standard by law). In contrast, 58 per cent of father-students took no leave, and only 16 per cent of men took leave longer than a month. We conclude that giving birth translates into a substantial delay in the PhD progress for mothers, creating a significant gap compared with their male colleagues.

Another aspect of significant gender difference, familiar from studies around the world, arises from looking into the private sphere of the families of the PhD students. We asked: "Who is responsible for most childcare work?" Of the male students, 67 per cent responded that they and their spouses carry the load equally, and only 5 per cent responded that the responsibility lies mainly on themselves. In contrast, of the women students, 43 per cent responded that they and their spouses share the load equally, and the other 57 per cent responded that the responsibility lies mainly on themselves. *Not even one of the women students said that her spouse is the main caregiver for the children.*

Two final issues where we identify gender-related differences are those of discrimination and of sexual harassment. We aimed to examine the issue of discrimination on a broad variety of possible backgrounds. We asked: "Have you ever felt discriminated against during your studies?" The gender difference here is very significant, as 67 per cent of women versus only 19 per cent of men reported that they have experienced discrimination. When asked on the grounds for the discrimination, 50 per cent of women mentioned gender discrimination and 19 per cent mentioned pregnancy or parenthood-based discrimination, while for men these issues were rarely mentioned. Another aspect of discrimination that was reported is age (17 per cent of women, 5 per cent of men). Other aspects – ethnic origin, religion, and social status – were mentioned by only very few respondents. See Table 6.2 for details.

On the issue of sexual harassment, we asked: "Did you experience sexual harassment during your academic studies?" To avoid any ambiguity, we provided the Israeli legal definition of sexual harassment (Knesset, 1998): (i) hostile atmosphere of sexual character in the organization, (ii) humiliating reference to a person on grounds of gender or sexual orientation, (iii) repeated references to a person focusing on his/her sexuality, (iv) repeated propositions of sexual nature, (v) lewd act, (vi) blackmailing a person to perform an act of sexual nature.

One of every five women (21 per cent), but only 2 per cent of men, reported that they experienced sexual harassment during their studies. See Table 6.3 for details. Among these, half of the women were harassed twice or more.

Only a minority of the women answered the question: "*By whom were you harassed?*" The answers included student colleagues, technicians, and lecturers, but none signalled their PhD advisor.

Table 6.2: Discrimination by background

	Women, *n=60*		Men, *n=207*	
	%	CI	%	CI
No	33	[30,36]	81	[78,84]
Yes (total)	67	[64,70]	19	[16,22]
Yes, by gender	50	[47,53]	1	[0,2]
Yes, By pregnancy/parenthood	19	[16,22]	2	[1,3]
Yes, by age	17	[15,19]	5	[3,7]

Note: Data presented by gender; CI stands for 95% confidence interval; bold indicates $p<0.05$.

Table 6.3: Experiences of sexual harassment. "Did you experience sexual harassment during your academic studies?"

	Women, *n=60*		Men, *n=207*	
	%	CI	%	CI
Never	79	[76,82]	98	[97,99]
Once	10	[8,12]	1	[0,2]
Twice or more	11	[9,13]	1	[0,2]

Note: Data are presented by gender; CI stands for 95% confidence interval; bold indicates $p<0.05$.

Summary and discussion

This research provides a broad, quantitative, and representative examination of the challenges facing physics PhD students in Israel. Its uniqueness is twofold. First, we surveyed the entire population of physics PhD students in Israel, rather than sample it. Second, we covered diverse topics of their academic and personal experience. The research explores a large variety of difficulties, starting with academic requirements and economic challenges, via personal difficulties related to family and parenthood, and including discrimination and harassment. Furthermore, in all these issues, we examine gender-related differences, and thus we learn about the distinctive experience of women physicists as a minority in a male-dominated field. Our findings confirm in part the findings of previous studies in other contexts.

The PhD students, both men and women, view the physics study course as one that is demanding, intensive, and difficult, and which requires overcoming professional, economic, and emotional challenges. From the academic aspect, there are many challenges that all students face, including the need to invest a great deal of time in their research, workload, frustration, and competitiveness. Because the curriculum demands do not allow for paid work, the students depend on a modest fellowship, which is significantly lower than their earning capacity in the labour market. The economic difficulty is exacerbated and stands out as a significant hurdle for married students. Moreover, coping with all of these also produces mental difficulties that manifest themselves in stress, anxiety, a continuing sense of uncertainty, and mental distress.

In addition to difficulties experienced by all students, women face additional hurdles. The gender-specific hurdles that we identified are hidden to the academic system, which is the reason why we call them 'glass hurdles'. The glass hurdles are of several types. First, women suffer more from psychological and physiological health problems. Second, women face challenges related to pregnancy, giving birth, and motherhood. Although the transition to parenting poses a significant challenge for both women and men, pregnancy and childbirth halt women's course of study much more. In addition, after giving birth, they carry a heavier burden of childcare and compared with male students after becoming fathers. Third, the findings show that while most women experienced discrimination during their studies, mainly on the grounds of gender, parenting, and family, most men did not experience discrimination at all. Furthermore, 1 in 5 women experienced sexual harassment during their studies (compared with a marginal rate among men). In this context, it is important to note that the literature points to serious and long-term effects of sexual harassment on women students and staff, as well as high attrition rates.

Our study shines a spotlight on the difficulties women in physics in Israel experience during their doctoral studies, and the additional hurdles they have to overcome in order to succeed in their academic careers. The additional difficulties women experience during their doctoral studies provide, however, only a partial explanation of the low proportion of women in physics graduate studies and academic staff. As we learned from previous research (Ceci and Williams, 2010; Buse et al, 2013; Lewis et al, 2016; Lewis et al, 2017), a number of cultural, social, environmental, and biological factors play a role in women's relatively lower representation in physics and other STEM fields. Given its persistence, the causes of gender disparity are likely to be complex and multiply determined, but our study reveals a set of gender-specific hurdles, which women need to negotiate to succeed in physics.

Academic institutes profess to believe, almost religiously, in the ideology of meritocracy, together with a liberal concept of equality, in their demands from the individual. Our research implies that, at the same time, these institutes are not taking care to level the playing field. They are either not aware of, or do not care about, the organizational climate in which women experience discrimination and sexual harassment, and do not deal with the gaps (in terms of time and attention that can be devoted to studies and research) that are generated between women students and their men colleagues, when the former become mothers.

Based on our findings, we suggest the following steps to be taken by academic institutes to remove the presently transparent 'glass hurdles' to women: addressing the problem of sexual harassment and promoting prevention programmes; promoting inclusive teaching and discrimination-free environments for women; adapting institutional policy to the special challenges that arise when combining PhD studies with family demands; raising the fellowship for students who are parents; and expanding the availability of psychological care for students in general and women students in particular. For related recommendations, see Directorate-General for Research and Innovation (2021).

At the same time, solutions to the situation are required that also refer to the change of accepted norms in physics. The current structure of the field, which obligates students to demanding and competitive studies, and the reality in which women experience a cold and discriminatory climate, must change – significantly, if the academy as a whole and physics departments in particular have a sincere desire – for a real change in gender imbalance.

Science in general and physics in particular will benefit from increasing the pool of talent and enhancing diversity. Therefore, we believe it is in the interest of the academic institutes and of the discipline of physics to increase the percentage of women at all stages of an academic career in physics. The currently transparent hurdles must first become visible to the academic system, and then be actively removed.

References

Aycock, L.M., Hazari, Z., Brewe, E., Clancy, K.B., Hodapp, T., and Goertzen, R.M. (2019) 'Sexual harassment reported by undergraduate female physicists', *Physical Review Physics Education Research*, 15(1): 010121. DOI: 10.1103/PhysRevPhysEducRes.15.010119

Bagilhole, B. and Goode, J. (2001) 'The contradiction of the myth of individual merit, and the reality of a patriarchal support system in academic careers: a feminist investigation', *European Journal of Women's Studies*, 8(2): 161–80.

Bailyn, L. (2003) 'Academic careers and gender equity: lessons learned from MIT 1', *Gender, Work and Organization*, 10(2): 137–53.

Barthelemy, R.S., McCormick, M., and Henderson, C. (2016) 'Gender discrimination in physics and astronomy: graduate student experiences of sexism and gender microaggressions', *Physical Review Physics Education Research*, 12(2): 1–14. DOI: 10.1103/PhysRevPhysEducRes.12.020119

Bondestam, F. and Lundqvist, M. (2020) 'Sexual harassment in higher education – a systematic review', *European Journal of Higher Education*, 10(4): 397–419. DOI: 10.1080/21568235.2020.1729833

Bostwick, V.K. and Weinberg, B.A. (2018) 'Nevertheless, she persisted. Gender peer effects in doctoral STEM programs' (No. w25028), *National Bureau of Economic Research*, Available from: www.nber.org/papers/w25028.pdf

Buse, K., Bilimoria, D., and Perelli, S. (2013) 'Why they stay: women persisting in US engineering careers', *Career Development International*, 18(2): 139–54. DOI: 10.1108/CDI-11-2012-0108

Cabrera, E.F. (2007) 'Opting out and opting in: understanding the complexities of women's career transitions', *Career Development International*, 12(3): 218–37. DOI: 10.1108/13620430710745872

Carmi, R. (2011) 'The team for examining the situation of women in the academic staff of high education institutes – report and recommendations', The Israeli Council for Higher Education (in Hebrew).

Ceci, S.J. and Williams, W.M. (2010) *The Mathematics of Sex: How Biology and Society Conspire to Limit Talented Women and Girls*, Oxford: Oxford University Press.

Clancy, K.B., Nelson, R.G., Rutherford, J.N., and Hinde, K. (2014) 'Survey of academic field experiences (SAFE): trainees report harassment and assault', *PLoS One*, 9(7): 1–9, e102172. DOI: 10.1371/journal.pone.0102172

DeVault, M.L. (1999) *Liberating Method: Feminism and Social Research*, Philadelphia, PA: Temple University Press.

Directorate-General for Research and Innovation (2021), *Horizon Europe guidance on gender equality plans*, Available from: https://data.europa.eu/doi/10.2777/876509

Epstein, C.F. (2004) 'Border crossings: the constraints of time norms in transgressions of gender and professional roles', in C.F. Epstein and A.L. Kalleberg (eds) *Fighting for Time: Shifting Boundaries of Work and Social Life*, New York: Russell Sage, pp 317–40.

Eran Jona, M. and Nir, Y. (2019) 'Women in physics in Israel: an overview', in G. Cochran, C. Singh, and N. Wilkin (eds) *Women in Physics: 6th IUPAP International Conference on Women in Physics Conference Proceedings: Vol. 2109*, American Institute of Physics.

Eran Jona, M. and Perez, G. (2021) 'Graduate student survey', work in progress.

Forster, N. (2001) 'A case study of women academics' views on equal opportunities, career prospects and work–family conflicts in a UK university', *Career Development International*, 6(1): 28–38. DOI: 10.1108/13620430110381016

Gerson, J.M. and Peiss, K. (1985) 'Boundaries, negotiation, consciousness: reconceptualizing gender relations', *Social Problems*, 32(4): 317–31.

Gofen, A. (2011) 'Academic career of graduates with excellent grades in SET fields 1995–2005'. Research report of the Federman School of Public Policy and Government, The Hebrew University of Jerusalem (in Hebrew).

Gonsalves, A.J., Danielsson, A., and Pettersson, H. (2016) 'Masculinities and experimental practices in physics: the view from three case studies', *Physical Review Physics Education Research*, 12(2): 020120. DOI: 10.1103/PhysRevPhysEducRes.12.0201209

Goulden, M., Mason, M.A., and Frasch, K. (2011) 'Keeping women in the science pipeline', *Annals of the American Academy of Political and Social Science*, 638(1): 141–62.

Guthrie, S., Lichten, C.A., Van Belle, J., Ball, S., Knack, A., and Hofman, J. (2018) 'Understanding mental health in the research environment: a rapid evidence assessment', *Rand Health Quarterly*, 7(3).

Harding, S. (1991) *Whose Science? Whose Knowledge? Thinking from Women's Lives*, Ithaca, NY: Cornell University Press.

Hargreaves, C.E., De Wilde, J.P., Juniper, B., and Walsh, E. (2014) 'Re-evaluating doctoral researchers' well-being: what has changed in five years?', London: Graduate School Imperial College London, Available from: www.imperial.ac.uk/media/imperial-college/study/graduate-school/public/well-being/Wellbeing-for-GS.pdf

Howe-Walsh, L. and Turnbull, S. (2016) 'Barriers to women leaders in academia: tales from science and technology', *Studies in Higher Education*, 43(3): 415–28.

Ivie, R. and Tesfaye, C.L. (2012) 'Women in physics: a tale of limits', *Physics Today*, 65(2): 47–50.

Johnson, P., Widnall, S., and Benya, F.F. (2018) *Sexual Harassment of Women: Climate, Culture, and Consequences in Academic Sciences, Engineering, and Medicine*, National Academies of Sciences, Engineering, and Medicine, Washington, DC: National Academies Press.

Kinman, G. and Wray, S. (2013) *Higher Stress: A Survey of Stress and Well-Being among Staff in Higher Education*, London: University and College Union.

Knesset, the Israeli Parliament (1998) 'Law for prevention of sexual harassment', Available from: www.knesset.gov.il/review/data/heb/law/kns14_harassment.pdf (in Hebrew).

Krumer-Nevo, M., Lavie-Ajayi, M., and Hacker, D. (eds) (2014) *Feminist Research Methodologies*, Bnei-brak: Hakibbuits Hameuchad (in Hebrew).

Lamont, M. and Molnár, V. (2002) 'The study of boundaries in the social sciences', *Annual Review of Sociology*, 28(1): 167–95.

Levecque, K., Anseel, F., De Beuckelaer, A., Van der Heyden, J., and Gisle, L. (2017) 'Work organization and mental health problems in PhD students', *Research Policy*, 46(4): 868–79.

Lewis, K.L., Stout, J.G., Pollock, S.J., Finkelstein, N.D., and Ito, T.A. (2016) 'Fitting in or opting out: a review of key social-psychological factors influencing a sense of belonging for women in physics', *Physical Review Physics Education Research*, 12(2): 020110.

Lewis, K.L., Stout, J.G., Finkelstein, N.D., Pollock, S.J., Miyake, A., Cohen, G.L., and Ito, T.A. (2017) 'Fitting in to move forward: belonging, gender, and persistence in the physical sciences, technology, engineering, and mathematics (pSTEM)', *Psychology of Women Quarterly*, 41(4): 420–36. DOI: 10.1177/0361684317720186

Libarkin, J. (2019) Academic Sexual Misconduct Database, Available from: https://academic-sexual-misconduct-database.org

Lorenz, K., Kirkner, A., and Mazar, L. (2019) 'Graduate student experiences with sexual harassment and academic and social (dis)engagement in higher education', *Journal of Women and Gender in Higher Education*, 12(2): 205–23. DOI: 10.1080/19407882.2018.1540994

Reinharz, S. and Davidman, L. (1992) *Feminist Methods in Social Research*, New York: Oxford University Press.

Sarseke, G. (2018) 'Under-representation of women in science: from educational, feminist and scientific views', *NASPA Journal about Women in Higher Education*, 11(1): 89–101. DOI: 10.1080/19407882.2017.1380049

Sekuła, P., Struzik, J., Krzaklewska, E., and Ciaputa, E. (2018) 'Gender dimensions of physics: a qualitative study from the European Research Area', Available from: https://genera-project.com/portia_web/Gender_Dimensions_of_Physics.pdf

Sue, D.W. (ed) (2010) *Microaggressions and Marginality: Manifestation, Dynamics, and Impact*, New Jersey: John Wiley & Sons.

Traweek, S. (1988) *Beamtimes and Lifetimes: The World of High-Energy Physicists*, Cambridge, MA: Harvard University Press.

Tytherleigh, M.Y., Webb, C., Cooper, C.L., and Ricketts, C. (2005) 'Occupational stress in UK higher education institutions: a comparative study of all staff categories', *Higher Education Research and Development*, 24(1): 41–61.

West, L.M. (2014) '"Something's gotta give": advanced-degree seeking women's experiences of sexism, role overload, and psychological distress', *NASPA Journal about Women in Higher Education*, 7(2): 226–43. DOI: 10.1515/njawhe-2014-0015.

PART III

European Initiatives

7

The GENERA Project: Experiences and Learnings of a Structural Change Project to Promote Gender Equality in Physics

Thomas Berghöfer, Helene Schiffbänker, and Lisa Kamlade

Introduction

The chapters in this collection evidence that physics is overwhelmingly a male-dominated research field. During the last decade, awareness of the gender imbalance which exists within all science, technology, engineering, and mathematics (STEM) fields has grown and, as a result, new EU initiatives have been developed to enhance inclusiveness and improve the gender balance in STEM fields. One of these initiatives has been the Gender Equality Network in Physics in the European Research Area (GENERA) project, which addresses this challenge through a 'bottom-up' approach for more gender equality in physics research organizations. In this chapter, we describe this project and present some major learnings for implementing gender equality in male-dominated disciplines. The aim is to provide insights into the implementation in practice and, by this, facilitate mutual learning for other research organizations.

This chapter is structured as follows. First, we describe the GENERA project and its activities towards establishing gender equality in physics research organizations. Second, we present experiences and lessons learned through the GENERA project. Finally, we formulate conclusions and recommendations for the implementation of projects to promote Gender Equality Plans (GEPs) in academic institutions in physics and other disciplines in STEM.

The GENERA project

The GENERA project was active between September 2015 and August 2018, and received funding from the European Commission under the Horizon 2020 call GERI.4.2014 'Support to research organization to implement gender equality plans'. This project started as a bottom-up initiative to systematically transform research organizations in physics for greater gender inclusiveness by implementing GEPs customized to the needs of the organizations. These GEPs should at least include the following:

- Conduct impact assessment/audit of procedures and include relevant data on Human Resources (HR) management, teaching, and research activities, in order to identify gender bias at the organization level.
- Implement innovative strategies to address gender bias; these should include family-friendly policies (for example, work schedule's flexibility; parental leave; mobility, dual-career couples); gender planning and budgeting; training on gender equality in HR management; developing gender dimensions in research content and programmes; and integrating gender studies in Higher Education Institution curricula.
- Set targets and monitor progress via indicators at the organization level.

The European Commission has given GEPs a very high priority in the implementation of an institutional change towards gender equality and a GEP is now mandatory for receiving funding from the EU. For the first time, research institutes were required to address gender equality in a systematic manner. The future will show whether GEPs are sustainable and effect the gender balance.

To maximize the impact and be as close as possible to the needs of the field, GENERA followed a bottom-up approach 'from physics for physics' considering the characteristics of physics research as a field to involve women and men physicists at all career levels in the design of GEPs. This was a new approach, as gender equality is traditionally an issue addressed by HR departments on the one hand or social scientists on the other hand. Furthermore, from the beginning, the idea was to embrace the high level of European/international networking in physics and to network all those interested in the topic of gender equality in physics.

The project activities were accompanied by an outreach campaign to inform and attract the physics community in Europe. During the lifetime of the project the starting consortium of 11 physics research organizations – and two professional support organizations, responsible for the project internal evaluation and outreach and valorization activities – could be extended by 23 additional physics research organizations. The additional organizations were integrated into the project activities as 'observers' without receiving additional funding from the project, which is in itself indicative of the relevance of the issue and the need for change.

The 11 organizations implementing GEPs stem from seven EU member states (and Switzerland) differ significantly in their structure, legislation, and current efforts. Three of the organizations came from Germany, two from Italy, and one each from France, the Netherlands, Spain, Poland, Romania, and Switzerland. Variations in country context are quite significant, with some organizations having to abide by national laws on gender equality and equality of opportunity, while such policies are not in place in others. As a prerequisite of the actual project activities these diverse starting conditions had to be understood in detail.

Each implementing partner of the project had to set up a local implementation team (LITeam) responsible for the organization's internal tasks associated with GEP implementation, which should include administrative as well as scientific staff ranging from lower (for example, students) to higher (for example, group leaders) hierarchy levels. This LITeam should be coordinated and linked to the project by an implementation manager (IM). The unity of a LITeam should ensure and symbolize the commitment and support of all staff as stakeholders to gender equality, which is essential in developing a successful and sustainable GEP (Verloo and QUING Consortium, 2012).

The entire process from preparing to implementing a GEP was accompanied by an evaluation conducted by a consortium member with expertise in gender and organizations, following a 'Critical Friend' approach (Balthasar, 2011). This means that the internal evaluator was in permanent exchange with the implementing organizations about progress, challenges, and problems, allowing to share a timely feedback with the consortium members and in consequence, allowing adjustment of the activities accordingly.

To ensure the long-term advancement of a structural change in physics research organizations and institutes and the establishment of gender dimensions in research content and programmes, the GENERA work programme included the preparation of a framework for the implementation of sustainable long-term monitoring of the effectiveness of GEPs. Furthermore, it was intended to make the project results available to the entire physics community in Europe.

The project was divided into three phases, as follows: the initial first phase was to identify the status quo and create a knowledge base; the second phase aimed to train the project team for the tasks ahead, develop all necessary tools including a scheme for taking relevant data on the organizational level and prepare the customization of GEPs in the participating organizations; and the final phase was to implement these GEPs and prepare suitable impact monitoring beyond the lifetime of the project. Structured according to these three phases, the following section describes the experience and lessons achieved during the GENERA project.

GENERA project: experience and practices

Status quo and preparation of a knowledge base

During the last 20 years, the gender imbalance in physics and neighbouring fields of science has been researched in multiple ways. Statistical studies (for example, European Commission, 2019) have confirmed that the number of women physicists is increasing in European countries in recent decades, but at a very slow rate. In addition to the overall lower numbers of women physicists in the research system, the statistics also show that with increasing career level the percentage of women decreases. This observation is described in the literature as a 'leaky pipeline' or 'glass ceiling'. However, physics and other maths-intensive fields also show an under-representation of women already exists at the entry point, at student level.

In order to derive an overview of the gender equality in physics research, GENERA started by carrying out a comprehensive literature review to understand causes of gender inequality as well as to map and identify successful gender-equality measures and conditions for improving research culture and environment in the fields linked to physics. More than 120 scientific publications released during the last four decades were reviewed by a team mainly of social scientists participating in the GENERA project. It became clear that these studies have received little to no attention from the physics scientific communities and organizations concerned.

To determine the current status of activities in the participating member countries, we introduced the initiative of 'Gender in Physics' Days (GiPDs). Subsequently, a series of 11 GiPDs were all over Europe. These one-day events helped draw a great deal of attention to the issue, but without any lasting effect. Although a lot of advertising was done for these events and especially men were invited to participate, the attendees were mainly women. The contributions of invited research-performing organizations and funding organizations made clear that while in all the participating countries there is a great interest in the topic, only limited and very rough sex-disaggregated data are collected which hardly allow evidence-based actions towards gender equality. The level of the discussions held during the GiPDs was an indication of whether the relevant stakeholders in a country (for example, France and Italy) already knew each other and whether the topic was already present or was being addressed for the first time.

In order to enable long-term monitoring of the impact of the GEPs, a scheme for collecting relevant data was developed. For this purpose, the participating organizations were asked which data they had available or could collect in the future. Unfortunately, the results turned out to be very sobering, as the defined data set consisted only of demographic information, work status, and incomplete information on career paths. It became apparent that there is a trend towards interdisciplinarity, and physics-oriented questions are often addressed by physicists in collaboration with scientists from other disciplines, making it very difficult for organizations to compile statistics for employed physicists only.

Moreover, the aim of GENERA was to take into account the interests of the scientists concerned and to obtain a bottom-up view as well as a top-down view on the status quo of gender equality in the participating institutions. During the project, a framework for conducting interviews with physicists – including a suitable questionnaire and interview guidelines – to assess career paths and working conditions in physics was established and tested with 85 interviews conducted with male and women physicists in partnering organizations. Despite selection and size of this sample limiting the statistical significance of the findings, this study confirmed the shift towards more and more scattered and precarious career pathways implying intensive mobility, short-term contracts, grant-funded positions, and the pressure of constant work. For women physicists this opens up many more potential holes in the 'leaky pipeline' and seriously challenges family–work reconciliation. While women-only networks and women-specific mentorship is given a high priority, both are rarely experienced by women physicists. The study revealed that there is indeed a growing awareness of gender imbalance in physics; however, there is insufficient knowledge about gender-equality measures and actions taken by the physics research institutions.

Overall, it became clear that it is very important to involve the scientists concerned to understand the challenges that physics organizations face when implementing gender-equality measures, in particular when they are starters on this journey. Interviews performed with management members – hard scientists often for decades – revealed that they found it difficult to understand the 'gender problem': as scientists used to measure and think in numbers and clear terms, it was difficult for them to understand that gender imbalance is constructed in organizational practices such as recruiting, promoting, or rewarding; and that assessment criteria are not gender-neutral as they are not equally adapted to the life circumstances of women; for example, the need to be internationally mobile may hinder women's ability to pursue a successful career in physics. Finally, the under-representation of women is not just in a given situation or the result of an individual choice, but systemic, constructed by the way physics is organized and by the field's culture. Here, an interesting, physics-specific aspect helped the project to progress: the interviews revealed that physicists are used to identifying problems and working on solutions as part of their professional identity. Thus, understanding the 'gender-equality problem' was a demanding step, while finding the commitment to solve the problem was not. Indeed, as soon as the need to implement gender-equality measures was clear, the question was how to find a solution.

Personnel and tools to customize and implement GEPs

As described earlier, IMs and LITeams had to be established by the implementing organizations. It was the decision of each organizations who to nominate as IM. In most cases, the person specifically hired for the project was assigned as IM. As a result, the team of IMs consisted mainly of young social scientists, physicists, and administrative employees at the beginning of their professional careers. The challenge of the project was to introduce and train this highly motivated team to the tasks both in their own organizations and within the project. Missing expertise and knowledge (for example, in organizational change processes and resistance) had to be complemented by targeted training within the project collaboration, which led to a heavy load on the IMs and delayed the start of the actual work. However, the more or less random, but well-mixed composition, of the IMs and the joint training they undertook together on the job turned out to have a rather positive effect on team building and underlined that support and practice structures are crucial to promote change (Dennissen et al, 2019). During the later cooperation, IMs supported each other in solving problems, dealing with resistance in their organizations, and instructing their LITeams.

To support the customization and implementation of GEPs in physics organizations, tailored instruments were developed: the GENERA Toolbox, GENERA Roadmap, and the GENERA Planning, Action, and Monitoring Tool (PAM-Tool).

The GENERA Toolbox (GENERA, 2018a) comprises known good practice in gender-equality measures that have so far been implemented by physics research organizations or in neighbouring research fields in Europe. More than 100 measures were collected from a variety of sources (surveys, experts, own experiences, and supplemented by literature review). These good-practice examples were categorized by fields of action:

- structural integration of gender equality,
- engaging leadership,
- flexibility, time, and work life,
- presence and visibility,
- gender-inclusive/gender-sensitive organizational culture,
- gender dimension in research and education,
- objectives and target groups (pupils; students; PhD students; PhD candidates and research assistants; postdocs and mid-career scientific personnel; professors; management and leadership).

Furthermore, all measures were classified according to the difficulty of implementation.

The GENERA Roadmap (GENERA, 2017) reflects models used in theories of organizational change and change processes of which the proposed steps are only a recommendation for a possible effective change (Krüger, 2006; Jann and Wegrich, 2014; Kotter International, 2017). It describes in detail the process leading to a sustainable GEP in the following six distinct steps:

- understand the structures of the organization and its rules (Learn),
- analyse the state of affairs using gender indicators (Analyse),
- design a tailored GEP with defined specific aims and measures to reach them (Design),
- implement the GEP and its measures (Implement),
- monitor the progress and adjust specific aims and measures (Monitor and Adjust),
- based on the results, adjust the GEP and move forward (Final Evaluation).

To help manage the structural complexity of the implementation activities, the evaluation team developed the GENERA PAM-Tool, which links measures and targets. The tool, developed throughout the accompanied evaluation process, consolidates experiences and expertise acquired throughout the project runtime. The development of the tool was aligned with the needs of the IMs and the (top) management of research organizations. It provides orientation, systematization, and causalities and serves as guidance in the GEP design process – from the first idea to the final GEP. In particular, it helps to answer questions such as (i) Which measures should be selected for specific targets? (ii) Which indicators can demonstrate the effects of a specific measure? Figure 7.1 provides a visualization of the PAM-Tool's action tree that starts from the three European Research Area (ERA) targets for gender equality defined by the European Commission for the ERA, addresses specific targets, and leads to tailored measures covered by the GENERA Toolbox. The PAM-Tool and its full description is available online (Schiffbänker and Hafellner, 2018).

GEP implementation and follow-up

Equipped with the developed tools, the duty of the IMs was to work out an organization's GEP with best-fitting measures adapted to the organization's needs. In regular meetings, the IMs were prepared step by step for the next tasks. These meetings were also used to capture the status of the implementation process in the individual implementing organizations and to provide an opportunity for IMs to share their experiences, successes, or

Figure 7.1: The GENERA PAM-Tool for planning and monitoring gender-equality plans in physics

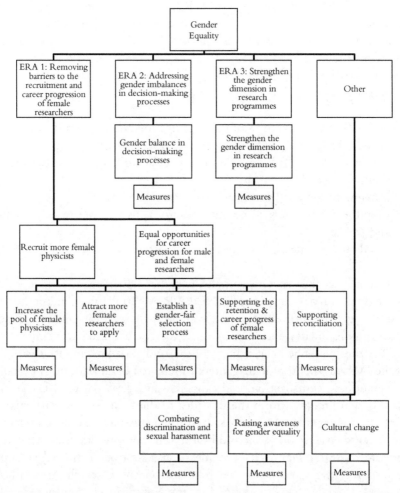

Source: Schiffbänker and Hafellner (2018).

failures. This procedure has been described by Bilimoria and Liang (2012: 13) as 'a network of change agents in peer organisations for the sharing of learning and best practices and support'.

By conducting interviews with the management and project leaders on site before and after GEP implementation, empirical evidence on the change process was provided by the evaluation team. This was completed by two online surveys sent between the interviews. Covering all implementing partners, about 100 interviews in the ex-ante phase and 70 interviews at the end of the project were conducted. The on-site interviews also had the purpose of determining the state of affairs and supporting the IMs in involving their management and leadership in the implementation process and pushing gender issues in the organizations further. To develop a collective understanding of gender equality and to foster organizational learning on this topic (Van den Brink, 2020) was a further intention.

In the frame of the project all implementing organizations managed to develop a GEP that met the original requirements and addressed the specifics of each organization. However, prolonged coordination efforts with stakeholders, internal organizational changes or changes in responsibilities, and extended approval processes meant that in some organizations the GEPs could not be officially adopted by the time the project was completed.

By extending the project consortium many additional physics research organizations could be integrated also in the discussion on how to sustain the project activities and widen the impact beyond the project lifetime. A Memorandum of Understanding (MoU) for establishing a network for collaboration in gender-equality policy for physics, 'The GENERA Network', was agreed. This MoU (GENERA, 2018b) describes objectives, membership, commitments, and organization of the network as well as the GENERA data set that members agree to annually provide.

Conclusions and recommendations

Within the GENERA project, a field-specific approach was followed to commonly address gender equality in physics. The project was successful in its main aim to enable physics research organizations to customize and implement GEPs. While all implementing partners developed a GEP during the project, not all GEPs had been signed by the respective managements before the project ended.

This development was facilitated by various factors.

- Interdisciplinary approach between 'field' people and social scientists: the bottom-up approach enabled the physics community on the organizational level (research organizations) as well as on the individual level (physicists) to really articulate their needs and challenges for more gender balance in the field. Having social scientists with gender expertise in the team as well as physicists actively engaged in various research topics in physics provided on the one hand a chance to identify field-specific (gendered) norms and on the other hand to bridge different discourses and to learn from each other as to how to make the physics field more gender-balanced and inclusive. It has proven to be very advantageous to actively involve the scientists concerned (here physicists) and include their approach and ability to address and solve problems for this task.
- Support is substantial – community of practice: the implementation of GEPs could succeed with young staff hired as IMs, supported by an experienced

LITeam, knowing the specifics of the organization, its power structures, and employees. Moreover, it was important to network the IMs in the project as a team so that they could share knowledge and support each other and also get through difficult and sometimes frustrating experiences, benefiting from an informal community of practice (Thomson et al, 2022). On the other hand, the IMs were able to acquire very broad expertise within the project, which qualifies them for further tasks in this domain.

Change takes time: the implementation phase has once again highlighted the difficulties associated with structural change processes in organizations and the necessity of support from management. The course of a GEP implementation is rather difficult to predict. Even when it is believed that a common understanding has been reached with all stakeholders, difficulties can still arise in the interaction of these forces. Changes in responsibilities, especially at the management level, can also lead to a fundamental questioning of the implementation of a GEP, so that all activities have to start again from the beginning and slow down change progress (Van den Brink, 2020).

• Using field specific attitudes for GEP design: interviews with physicists on all career levels and the management performed by the evaluation team provided good insights into the challenges that physics organizations face when implementing gender-equality measures, in particular when they are starters on this journey. The interviews revealed that physicists are used to identifying problems and working on solutions as part of their professional identity. Thus understanding the 'gender-equality problem' was a demanding step, while finding commitment to solve the problem was not. How to become active for more gender balance in physics and what to do concretely to obtain intended results, was of interest for (top) management. But there was a lack of experience on how to really decide on the next steps and what the effects and impacts of these steps would be. This was well summarized by one interviewee from the top-level management: "Tell me what button to press and what then is the outcome." This quote illustrates that there is no understanding yet about the complex and systemic way to implement measures and that outcomes cannot be predicted precisely, but depend a lot on the context and how they are embedded in other activities.

• In-time Feedback Loops: the decision to accompany the project by a continuous Critical Friend evaluation has proved to be very advantageous. The continuous monitoring of the progress and reporting to all the actors involved made it possible to avoid undesirable developments in the project; for example, the evaluation team learned that some organizations faced difficulties to identify appropriate targets or select the right measures to achieve their targets. Accordingly, the team developed the Planning, Action, and Monitoring Tool (PAM-Tool) which links measures and targets and may be used in essentially all STEM fields with similar challenges concerning gender equality.

- Aiming for sustainability: a time-limited project can certainly not address and solve all existing problems that exist with regard to gender equality in science – in this case physics. GENERA has built on previous activities and contributed to the desired structural change regarding the implementation of gender equality in physics research organizations. The GENERA Network, which was founded as a result of the project, is intended to make the project results available to the entire physics community in Europe and to continue to pursue gender equality in physics together in a network. Interested research organizations are welcome to join the GENERA Network.

While providing some lessons here for change agents aiming to improve gender balance in other male-dominated fields, it should be taken into account that progress is never linear, that challenges and backlashes might emerge, and that outcomes are process-orientated (Janssens and Steyaert, 2019). We should be aware as well that a GEP is only a starting point, and that changing the culture in an organization or in a field requires continuous discussion in organizations to enable organizational learning (Van den Brink, 2020) in order to change processes and practices.

Beyond that, the focus in the near future needs to widened, to take into account other social categories as well, and aiming for a broader, intersectional inclusion (Woods et al, 2022) in various science fields.

Acknowledgements

The GENERA project was supported by the European Commission under grant number 665637.

References

Balthasar, A. (2011) 'Critical friend approach. Policy evaluation between methodological soundness, practical relevance, and transparency of the evaluation process', *German Policy Studies*, 7(3): 187–231.

Bilimoria, D. and Liang, X. (2012) *Gender Equity in Science and Engineering: Advancing Change in Higher Education*, New York: Routledge.

Dennissen, M.H.J., Benschop, Y.W.M., and Van den Brink, M.C.L. (2019) 'Diversity networks: networking for equality?', *British Journal of Management*, 30(4): 966–80. DOI: 10.1111/1467-8551.12321

European Commission (2019) 'She Figures 2018', Available from: https://ec.europa.eu/info/publications/she-figures-2018_en

GENERA consortium (2017) 'The GENERA Roadmap for the implementation of customized Gender Equality Plans', Available from: www.genera-network.eu/genera-roadmap

GENERA consortium (2018a) 'The GENERA Toolbox – developed by and for physicists', Available from: www.genera-network.eu/genera-toolbox

GENERA consortium (2018b) 'The Memorandum of Understanding for establishing a network for collaboration in gender equality policy for Physics', Available from: www.genera-network.eu/mou

Jann, W. and Wegrich, K. (2014) 'Phasenmodelle und politikprozesse: der policy cycle', in K. Schubert and N.C. Bandelow (eds) *Lehrbuch der Politikfeldanalyse*, Oldenbourg: De Gruyter.

Janssens, M. and Steyaert, C. (2019) 'A practice-based theory of diversity: respecifying (in) equality in organizations', *Academy of Management Review*, 44(3): 518–37.

Kotter International (2017) 'Kotter International', retrieved from '8 Steps to Accelerate Change', Available from: www.kotterinternational.com

Krüger, W. (2006) *Excellence in Change – Wege zur strategischen Erneuerung* (Vol. 3), Wiesbaden: GWV Fachverlage.

Schiffbänker, H. and Hafellner, S. (2018) 'The GENERA PAM (Planning – Action – Monitoring) Tool', Available from: www.genera-network.eu/pam:pam

Thomson, A., Palmen, R., Reidl, S., Barnard, S., Beranek, S., Dainty, A., and Hassan, A. (2022) 'Fostering collaborative approaches to gender equality interventions in higher education and research: the case of transnational and multi-institutional communities of practice', *Journal of Gender Studies*, 31(1): 36–54. DOI: 10.1080/09589236.2021.1935804

Van den Brink, M. (2020) 'Reinventing the wheel over and over again: organizational learning, memory and forgetting in doing diversity work', *Equality, Diversity and Inclusion*, 39(4): 379–93. DOI: 10.1108/EDI-10-2019-0249

Verloo, M. and QUING Consortium (2012) *Final QUING Report*, Vienna: Institute for Human Sciences.

Woods, D.R., Benschop, Y.W.M., and Van den Brink, M.C.L. (2022) 'What is intersectional equality? A definition and goal of equality for organizations', *Gender, Work and Organization*, 29(1): 92–109. DOI: 10.1111/gwao.12760

8

Women in String Theory

Yolanda Lozano and Marika Taylor

Introduction

This chapter discusses the origins and impact of a European network on string theory and gender. The authors of this chapter are string theorists who have led diversity initiatives within this research field and the chapter draws on the author's personal experiences of and knowledge about the string theory community.

String theory is a field of theoretical physics research that aims to create a unified theory of all the interactions in Nature bringing together particle physics and Einstein's theory of gravity. Of all the research fields in physics, string theory is the most disconnected from experimental observations: current day experiments cannot test string theory, just as the gravitational waves predicted by Einstein's 1915 theory of gravity were not observable until the beginning of the 21st century.

Progress in string theory is driven by conceptual theoretical developments rather than by experimental discoveries. Judgements about the quality and importance of research in string theory are arguably more strongly affected by culture and sociological factors than they are in fields of research in physics that are more closely connected to experiments. String theory is perhaps the paradigm of theoretical research, driven by ideas that are perceived to be 'breakthrough' and 'brilliant'. Since ideas cannot be tested with experiments, the importance of a scientific paper is subjective, and its impact may be influenced by the reputations and networks of the authors.

While academic posts are competitive across the physical sciences, competition for positions in string theory is exceptionally high, from doctoral study through to permanent academic positions. String theory is a proposed physical 'theory of everything', the frontier of research, and attracts many high-achieving researchers. However, the number of posts is relatively small as most of the research in physics departments focuses on experimentally accessible science. String theory groups across Europe use a joint application system for postdoctoral positions and receive well over 400 applications for around 40 positions each year. A permanent academic post in string theory might well receive 200 or more applications from highly qualified researchers. Researchers typically need at least four to six years of international postdoctoral experience to secure a permanent academic post, and thus the postdoctoral research phase is extended and uncertain.

Within the string theory community, the folklore was that around ten per cent of the research community were women, but no explicit data had been collated and analysed until the European project of 2013–17 described in this chapter. While various physics professional bodies, scientific research councils, and universities initiated the collection of gender data earlier, this type of data aggregates all research fields in physics and astronomy, potentially concealing significant variations between different research fields in physics.

Other chapters within this volume explore sociological and cultural factors that affect women in physics. String theory is a field of research within physics where these factors are exhibited particularly dramatically. Further, not only is the importance of work in string highly subjective but it is also influenced by the geographic location of the researcher and their standing within global research networks.

Feelings of isolation are a well-known barrier for women in physics. Theoretical physics research groups are small relative to experimental groups and women researchers are often the only woman in their research group, right through from doctoral research to permanent post. Accordingly women researchers in string theory usually have to reach out beyond their own institutions to find other women peers and mentors, while men can more easily access such support within their own research environments. Lack of peer and mentor support are key factors in women deciding to leave scientific research.

String theory publications are usually written in small collaborations of at most three or four researchers, and women researchers will typically be the only woman in the collaboration. Research collaborations are dynamical, forming quickly to work on a new idea and then dissolving once the set of publications on this theme is complete. These dynamical collaborations are key to successful and impactful research but there are considerable barriers to women establishing collaborations. The under-representation of women makes it harder for women to network. Collaborations are often formed outside formal working environments, in settings where women may not have been invited.

String theorists are expected to travel extensively, not only to raise their visibility and to network but also to form collaborations. Important collaborations are often formed during research conferences and extended research visits to other institutions. This culture creates major barriers for those who cannot travel extensively due to family and caring responsibilities. Very few women hold leadership roles within the field of string theory. Where they do have leadership roles on committees or selection panels, they are often the only woman, thus making it harder for them to confront biases or to promote gender equality.

Clearly all these factors mean that women string theorists face considerable isolation throughout their scientific career. This isolation may explain why women string theorists did not openly discuss or confront gender inequality until the EU-funded project described in this chapter. The idea to coordinate women string theorists across Europe to address gender equality came out of discussions between women researchers at conferences as they shared personal experiences and realized there were common themes. One consequence of the isolation of women researchers was that even high-achieving women string theorists would attribute barriers or challenges they encountered to their own weaknesses and faults rather than systemic gender equality. Discussions between researchers at conferences highlighted the commonality of issues and stimulated the women research community to take action.

Formation of a European network

In 2012 a group of European string theorists led by Yolando Lozano and Silvia Penati succeeded in obtaining funding from the EU FP7 programme for a European Cooperation in Science and Technology (COST) action in string theory that combined goals in both theoretical physics and gender equality. The COST action 'The String Theory Universe' ran from 2013 to 2017 (details of the programme can be found in COST, 2022). The network was originally initiated by 30 women researchers from across Europe but under FP7 rules the COST action was open to all European and Israeli groups. By 2015 most string theory research groups had joined and the network had 550 members, of which 87 per cent were men.

The main goal of the COST action was to carry out state-of-the-art research in the broad field of string theory, by exploiting and promoting complementary expertise of different research groups across Europe. The action also had the goal of stimulating collaborations with neighbouring research areas in physics and mathematics. Together with these scientific goals, the network aimed to develop specific gender actions and to support active participation of women scientists. The outcome of the COST action was therefore expected to have a positive impact on both science and society at a European level, in line with the strategic priorities of the EU FP7 programme.

The gender aspect of the COST network had four main goals: to increase the visibility of women string theorists and their participation in leadership roles; to increase knowledge about gender issues within the string theory community; to collect quantitative and qualitative data on gender; and to identify and implement gender-equality measures. The community had almost no data on its own demographics, so a key goal for quantitative data collection was to understand the gender structure of the community. COST was also keen to initiate the study of qualitative data via approaches such as culture surveys and focus groups.

Since the COST network was founded and led by women string theorists, women immediately became more visible within the European research community. In fact, not only were the two leaders of the network women but also 58 per cent of leadership roles were held by women. Prior to this network, women string theorists had rarely been included in leadership roles such as European network leads or conference advisory committees. As we discuss in the conclusions, the European string theory community now routinely includes women in such roles in the wake of the COST action.

Before 'The String Theory Universe', equality issues were not discussed openly within the string theory community. Some researchers within the community had awareness and knowledge of studies and research on gender in science, particularly women researchers. However, much of the research community had not been exposed to any research on gender in science. Women in the string theory community felt that many colleagues were unaware of challenges and barriers faced by women scientists.

The COST network developed a programme of activities aimed at increasing awareness of gender issues. Many of these activities were carried out through annual workshops on 'String Theory and Gender'. These workshops combined scientific talks on string theory with invited talks from gender experts, showcasing academic research related to gender in science. These workshops attracted increasing numbers of participants from the string theory community as the COST network developed. While many workshop participants were women, the broader impact of these workshops was gradually acknowledged by everyone within the community. The network leaders also attended events such as the annual 'Gender Summits' (Gender Summits, 2022), to build connections with other communities working on gender equality in science.

The COST network also organized activities for those with little awareness of gender issues within plenary sessions of string theory conferences. Activities included talks by women scientists and policy makers presenting gender data and discussing topics such as implicit bias and microaggressions. Talks on gender issues were followed by discussions and debates in which all participants were encouraged to participate. Women string theorists felt that it was important for their colleagues to be made aware of their experiences using constructive and non-confrontational approaches. An important activity in this direction was the performance of a play at the final conference of the network. This play was written by a professional scriptwriter, working in consultation with women researchers, and it exhibited real incidents of sexist microaggressions and discrimination experienced by women in the field (and similar to those discussed in Chapter 5). We will explore these experiences later.

All of the network activities were developed bearing in mind that it can take more than data or academic research on gender to convince scientists that there are gender inequalities in their own working environment. Scientists may not reflect on the environmental and sociological factors that influence women's decisions but may instead genuinely believe that women do not enter or stay in the research field just because of lack of interest. Discussion of the personal experiences of women researchers and the barriers they face is essential for progress.

The philosophy of the COST network was that all scientists have an inherent responsibility to identify and address inequalities, particularly in their roles as the educators of the next generation of scientists. As the survey data discussed later in this chapter shows, some scientists feel that gender equality is affected by a wide range of societal, educational, and cultural factors, and their own influence on these wide issues is limited. However, the network leaders felt that this argument should not be used to detract from the responsibility of all scientists to address issues within their own community.

Researching gender in string theory

One of the core goals of the COST network was to explore gender inequalities within the string theory community. Prior to the creation of the COST network, string theorists in Europe understood that approximately 10 per cent of their research community was women, but no systematic data had been collected or analysed. The COST network initiated the study of diversity data in string theory.

Since the primary goal of the COST action was research in string theory, almost all permanent staff in string theory in Europe joined the action to participate in events organized by the network. The membership of the action confirmed the anticipated gender balance: 11 per cent of the permanent staff in the network were women. Among postdoctoral researchers and doctoral students, the percentages were 9 per cent and 20 per cent respectively.

These figures suggest that there is a 'leaky pipeline' in the transition between doctoral study and postdoctoral research. However, the COST network's figures for early career researchers need to be read cautiously. The percentage for postdoctoral researchers is broadly in line with the postdoctoral appointments made within Europe over this period, as we discuss later. The doctoral student percentage does not seem to be as representative of the doctoral researcher community in string theory in Europe. Nearly 300 permanent staff were members of the action but only 134 doctoral students were members. Most senior researchers in string theory have one or more doctoral students suggesting that many doctoral students may not have been members of the action. Survey evidence suggests that the representation of women in string theory at doctoral level is around 12–15 per cent, which is still higher than the postdoctoral researcher percentage.

The COST network aimed to explore the reasons for the fraction of women dropping between doctoral study and postdoctoral research. Possible issues could include women being less likely to apply for postdoctoral positions and women having lower success rates when they did apply. The network analysed quantitative data and later used survey methods to understand cultural factors that could affect the retention of women.

Independent evidence of the gender balance of the postdoctoral string theory community was obtained from the records of the European joint postdoctoral application portal coordinated by Antoine van Proeyen in Leuven (Joint Postdoc Applications, 2022). Applicants apply in the autumn of each year with all offers of postdoctoral positions being made in a coordinated manner by groups across Europe in December and January. From 2005 onwards the application portal has collected information about the gender of applicants. The total number of annual applications to this portal has increased from around 200 in 2005 to nearly 500 in 2020 but the percentage of women applicants has changed very little during this period. The aggregated percentage of women applicants is 10.7 per cent, with the annual fraction varying between 8 per cent and 13 per cent, and there has been no statistically significant change since 2005. (Applicants can now choose a non-binary gender but this was not an option in the earlier years of the portal; the percentage of applicants registering a non-binary identity is less than 1 per cent.)

Applications to the European joint postdoctoral application portal are made by graduating doctoral students and postdoctoral researchers from Europe as well as the rest of the world. One thus cannot directly compare diversity data from the application portal with that of the COST network on researchers within Europe. However, the postdoctoral application data not only confirm the anticipated figure of around ten per cent women but also highlight the lack of change since 2005: the fraction of women researchers is not increasing. Many scientific researchers consider the low representation of women in permanent posts to be a relic of gender issues in previous decades and believe that the current pipeline of early career women researchers is larger. The European joint postdoctoral application portal data demonstrates that gender issues are not just historical in string theory; the gender balance of the pipeline is not improving.

The aggregated success rate for European postdoctoral applications is around 9 per cent. Due to the low numbers involved the success rate for women varies considerably between years but the aggregated success rate for women is 8.9 per cent, which is comparable to that for men (9.6 per cent). The aggregated fraction of women awarded postdoctoral positions is around 10 per cent, which is broadly in line with the 9 per cent fraction of women postdoctoral researchers recorded by the COST network.

There are possible explanations for the latter fraction being slightly lower. The European application portal only tracks postdoctoral positions funded by research grants but some early career researchers in string theory are funded by individual research fellowships. Fellowships include the prestigious EU Marie Sklodowska-Curie fellowships as well as fellowships funded by national scientific research councils throughout Europe. Lower application rates and success rates for these fellowships would have the consequence of reducing the fraction of women in the string theory postdoctoral community. Moreover, gender imbalances in early career fellowships could then become amplified in long-term and permanent track fellowships through the well-known 'Matthew effect', whereby pre-existing advantages/disadvantages become entrenched over time (see, for example, Bol et al, 2018).

In recent years, funding bodies around Europe have analysed the gender split of applications and awards in considerable depth. Some research councils in Europe have indeed reported lower success rates in fellowships for women but in most European countries the success rate is similar for men and women (Swedish Research Council, 2020; Gender Net Plus EU, 2021; Irish Research Council, 2022). Note, however, that most funding bodies report application and success rates aggregated over scientific disciplines which obscures the low application rates of women within fields such as theoretical physics. The recent policy brief (Gender Net Plus EU, 2022) highlights the importance of breaking data down by discipline, particularly for disciplines where gender disparities are large.

Although the success rates for men and women are similar, a number of analyses have noted that the application rates from women are lower. For example, the UK collects extensive statistical data on staff working in higher education and reports by the UK funding bodies note that the representation of women in fellowship applications is lower than the representation of eligible women within higher education (Royal Society London, 2020; UKRI, 2020).

The quantitative data from the European postdoctoral application portal give important insights into the leaky pipeline. While women applicants have comparable success rates to men, the fraction of applicants who are women has been stagnant at around 11 per cent since 2005. This percentage of 11 per cent is lower than the percentage of women doctoral researchers on the COST network suggesting that women doctoral students may be less likely to apply for postdoctoral positions than men. This seems in line with the observations made by the research funding bodies with respect to fellowship applications.

The European string theory community does not have a common portal for doctoral applications, nor does it store collated data about the doctoral research community demographics. Data collected from individual countries and research councils (GenHET, 2022; UKRI, 2022), however, suggests that around 12–15 per cent of doctoral researchers in string theory are women. As we discuss in the conclusions, it would be interesting to collate and explore further data that characterizes the transition between doctoral study and postdoctoral research within the string theory community.

The survey

As the COST action progressed, gender issues were increasingly discussed in conferences, not just those devoted specifically to gender in string theory, but also in mainstream conferences on string theory. From these discussions it was apparent that the string theory community held a range of views on the gender imbalance in string theory and on measures to address it. The COST network thus decided to carry out a survey to explore the views of the research community.

This survey was carried out at the end of the COST action in 2017 as earlier attempts to survey the research community had very low response rates. The higher response rate for the final survey could itself be viewed as a success of the COST network, as more researchers were willing to engage with discussions on gender issues by the end of the action. In addition, the data acquired from the 2017 survey could be used as baseline data for exploring changes within the string theory community in future.

The format of the survey was an online anonymous questionnaire consisting of 15 core questions. Each of the questions was posed in the form of a statement, to which the respondent could strongly agree, agree, neither agree nor disagree, disagree, or strongly disagree. Respondents could also add free-form comments explaining their responses. The statements were grouped around five themes: equal opportunities (EDI); family and caring responsibilities; gender and work modalities; gender initiatives within the COST action; and gender actions for the future. Phrasing was chosen so that the meaning of the question was clear to respondents from all around Europe; use of specialist equality and diversity terminology was avoided. Questions on the gender and professional status of the respondents were included for interpretation of the survey results. The open questions within the survey data formed the first qualitative data collected within the string theory community and give important insights into the culture and environment.

The full survey data can be found here (COST MP1210). A total of 172 people participated, of whom 112 were men and 50 were women. One respondent indicated their gender to be 'other' while nine preferred not to state their gender. Fifteen of the participants were doctoral students, 44 were postdoctoral researchers, and 113 were permanent researchers.

The first theme within the survey was *equality of opportunity*. Respondents were asked whether men and women have equal opportunities for career advancement, whether men and women are treated equally within their university departments, and whether they consider the string theory research environment to be particularly difficult for women relative to other science and engineering disciplines. A significant difference in perceptions between men and women respondents was observed for all three questions.

Among men respondents, 58 per cent felt that women and men have equal career opportunities while 31 per cent disagreed. Some men respondents felt that women are actually advantaged because of the policies supporting women in science. The situation was reversed for women respondents: only 27 per cent agreed that women and men have equal career opportunities while 57 per cent disagreed. Many of the narrative comments noted that while women and men may in principle be treated equally there are substantial differences in practice due to conscious and unconscious biases; impacts of pregnancy and the unequal distribution of caring responsibilities.

Around 80 per cent of men felt that staff are treated equally in their own departments regardless of gender; only 11 per cent of men felt that there were gendered inequalities. There was again substantial difference between the perceptions of men and women researchers. Only 39 per cent of women felt that staff were treated the same regardless of gender while 27 per cent disagreed. Some men noted specific examples of inequalities while a number of women raised cultural issues associated with their working environments being dominated by men.

String theory is perceived to be a particularly competitive field in physics. Nonetheless only 24 per cent of men agreed that the string theory scientific environment is particularly difficult for women compared with other scientific disciplines; 40 per cent of men disagreed with this statement. The responses from women were different: 49 per cent of women agreed that the environment is particularly difficult while 20 per cent disagreed.

Narrative comments referenced the unique nature of string theory research within physics. The lack of observational data in string theory is perceived to amplify biases with the importance of scientific work being subjective and its impact within the research community being affected by factors beyond scientific quality. String theory is a global research community and respondents noted that the need for an extended period of international mobility during the postdoctoral phase conflicts with caring responsibilities and family life.

The survey included five questions related to *family and caring responsibilities*. Three of these related to whether women and men researchers with caring responsibilities are disadvantaged within their careers. The other two questions related to institutional and national policies on childcare provision and parental leave.

A large majority (over 80 per cent) of respondents of both genders agreed that women with caring responsibilities are disadvantaged in their careers. While respondents felt that this is not unique to string theory, the specific context of the research environment was felt to amplify the problems. Comments again noted the need for an extended period of international mobility as a postdoctoral researcher as well as the fact that many permanent posts are tenure track so women researchers are under pressure to meet tenure requirements during their childbearing years. Respondents also highlighted well-known impacts of caring responsibilities on women. For example, some felt that the impacts of parental leave on publication records are not adequately taken into account in the evaluation of research grant applications.

The survey also asked whether men with young families are disadvantaged. 45 per cent of men agreed with this statement while only 33 per cent of women agreed. Women respondents noted that women with young children are almost always the primary carers while for men with young children this is usually or often not the case.

Moving to policies, respondents were asked whether their institution and country give adequate support for childcare. Perhaps unsurprisingly a substantial number of men and women respondents felt that there is not adequate support: 42 per cent of men and 45 per cent of women. Respondents interpreted the question both in terms of the availability and cost of childcare, as well as parental leave policies. These vary considerably around Europe but a common theme in responses was that the differences between maternity and paternity leave often force women to take a higher share of parental leave. A majority of respondents (over 60 per cent) agreed that national or institutional changes to parental leave policies such as women and men having equal parental leaves could have positive impacts on women researchers.

The survey asked two questions about the *modality and research environment* in string theory. Respondents were asked whether the high international mobility required to build a career in string theory interferes with family life; 85 per cent of men and 92 per cent of women agreed with this statement. However, the comments note that moving between research groups in different countries is a key part of early career training and networking. Respondents do not expect this culture to change and researchers who cannot move are inherently disadvantaged in terms of visibility within the field and reduced opportunities for networking and for forming collaborations.

The second research environment question related to the sensitive issue of harassment: were respondents familiar with their institution's policies on harassment and would they know how to proceed if a junior colleague wanted to raise a complaint? About 43 per cent of men and 39 per cent of women were aware of policies and how to raise complaints. However, respondents felt that harassment and other misconduct is very difficult to confront; making complaints could harm their career progression. Some respondents also felt that the general awareness of conduct and harassment policies within their departments was low.

The next group of questions explored viewpoints on *gender initiatives* within the string theory community. Respondents were asked whether they considered that the COST network had contributed significantly to addressing gender equalities in the string theory community: 58 per cent of men respondents and 71 per cent of women respondents agreed that it had. Some respondents felt that the key achievement of the COST network was to initiate the dialogue about gender and diversity issues within string theory. The string theory community within Europe was felt to have much more awareness about factors that affect under-represented groups following the activities of the COST network.

Respondents were also asked whether the conferences dedicated specifically to gender and string theory had had a good impact and should be continued. About 50 per cent of men respondents and 67 per cent of women respondents agreed with this proposition. Many people said that they had not themselves participated in these meetings but nevertheless believed that these had been useful and should be continued.

String theorists travel extensively to network and to form collaborations. Researchers at all career stages are expected to participate in workshops of three or four weeks' duration at institutes such as the Aspen Center for Physics and the Kavli Institute for Theoretical Physics in Santa Barbara. Some prestigious workshops only allow participants to attend if they can stay for at least three weeks. Family and caring responsibility constraints on travel can have substantial impacts both on the visibility of a string theorist and on their ability to work on cutting-edge topics.

One of the survey questions probed this topic by asking whether respondents felt that more effort should be put into the organization of conferences and workshops with respect to provision of childcare, as well as increased flexibility in lengths of stay for workshops. There was agreement among respondents that this is an important issue in string theorists: 64 per cent of men respondents and 72 per cent of women respondents agreed with the proposition. However, respondents also noted that addressing these issues is challenging. Conference organizers have no funding for childcare and parents can often not bring children to conferences or extended workshops due to school constraints. Increasing the flexibility about lengths of stay for workshops is possible and this has been implemented by some research institutes. However, participants may not get the full networking and collaboration benefits of workshops if they can only attend for a short period.

Respondents were asked whether they themselves undertake activities to address diversity issues for example mentoring or outreach. Only 34 per cent of men respondents do so but 72 per cent of women respondents are involved in gender-related activities. Work to address diversity issues thus seems to fall disproportionately onto the under-represented group itself, as has been seen in many other contexts. It is common for researchers to feel pressure to use their time for scientific work rather than for diversity activities than are less valued by their research communities. However, some respondents phrased these concerns even more strongly: they felt that active involvement in gender activities could be viewed negatively by colleagues and could harm their career progression.

The final question on the survey related to *future initiatives on gender* within the string theory community. A recurrent theme throughout the COST action was that some researchers felt gender issues could not be impacted significantly by actions of a single research community, with larger societal changes needed for any major impacts. To gauge whether this feeling was widespread, respondents were asked to choose one activity on which the community should focus. The choices were supporting women researchers (through mentoring, leadership training, and so on); working with the mostly male research community to understand implicit bias and other diversity issues, or focusing diversity efforts outside the string theory community (for example, working with schools and policy makers).

Opinions were split. About a quarter of respondents of both genders felt that the support should be focused on women researchers within the string theory community. Only 19 per cent of men favoured the approach of working with men in the community to improve their understanding of diversity issues, but 37 per cent of women respondents chose this option. Around 46 per cent of men and 29 per cent of women favoured focusing efforts outside the immediate string theory community that is working on outreach and with policy makers. Thus almost half of men put the emphasis on actions within the wider society whereas a majority of women felt that significant positive changes could be made within the string theory community itself, either by supporting women or by addressing biases and issues that impact on under-represented groups.

Experiences of sexist microaggressions

In this section we explore the experiences of microaggression, sexism, and discrimination that around 100 women reported to the COST action when the script for the theatre play was being written. The following is an excerpt from the letter sent to women researchers asking for them to contribute their experiences:

It is about situations happening in everyday life that, like a little but constant drop, end up making a hole in women's self-esteem or embittering to a certain degree the relation with their colleagues. They are just comments or attitudes that are repeated and transferred by imitation. Bringing them to the light could help stop them. As with unconscious bias, sometimes the situation barely appears in the fringe of women's own consciousness; sometimes only an uneasiness is felt and the woman 'learns' that she is not well adjusted.

Many common themes emerged in the experiences reported by women string theorists. Women often feel pigeonholed into administrative roles within their departments and even within their scientific collaborations. Even worse, women researchers are often *assumed* to be administrators in their departments rather than scientists. Even when it is known they are not administrators, it may be assumed that women researchers can take care of administrative tasks. Within collaborations women researchers may be expected to write up papers, almost as if they are the designated secretary for the collaboration.

Many reported experiences link to low status and lack of respect for professional expertise. Women researchers find that their contributions to discussions are often ignored but if the same ideas are repeated by men they are taken much more seriously. Sometimes women even have their own ideas explained back to them in patronizing tones. Questions from women in conferences or discussions may be dismissed, sometimes with a joke or a laugh. Women conference session chairs report that their instructions to speakers or to members of the audience are often disregarded and over-ridden.

Women feel they are not given sufficient credit for their work. Papers may be verbally cited in conferences by the names of the men in the collaboration, omitting women authors. When women present their work, members of the audience may address their questions to men and indeed men may directly answer questions addressed to their women colleagues. The independence of women researchers is questioned even when they are experienced researchers. It may be assumed that the woman is not the leader of a collaboration, even when the woman researcher has led work on this topic for some time. Senior women are asked about working with their doctoral supervisors even when they have not been worked with their supervisors for many years. Women string theorists notice that colleagues express considerable surprise about them being appointed to senior roles such as tenure track or permanent posts. Implicit bias is manifest in some recommendation letters for early career researchers and certain doctoral student supervisors seem to rate male students higher by default.

Women string theorists find it hard to raise diversity issues within their working environments. Women commonly hear complaints about gender initiatives from their colleagues. Colleagues may complain about requirements for 'politically correct' gender-neutral language or other modern working practices. When diversity issues are raised in committee meetings, colleagues often equate gender equality with lowering academic standards. Women who express views on diversity or engage in diversity initiatives experience criticism, particularly about the time spent on these activities instead of on research.

Women regularly experience inappropriate behaviours from colleagues or students and these behaviours are not challenged by their male colleagues. Examples include students being disrespectful or even offensive without consequences, and colleagues and students scrutinizing and criticizing the clothes and appearance of women. Men may reference discriminatory stereotypes, such as women who are good at physics having 'male brains'.

Women also reported more serious incidents of direct discrimination and even harassment. However, while women would talk about having experienced harassment in private, they found their experiences too frightening and painful to have these shared in public via a play.

When the play bringing together these experiences was performed at the final COST conference it had a strong impact on participants and provided considerable discussion. Men and women researchers again displayed very different perspectives on gender issues. Some men researchers felt that the situation for women would automatically improve with a younger generation coming through and no further actions for gender equality were needed. Most women researchers disagreed with what they considered to be an over-simplistic and over-optimistic view of deeply entrenched gender inequalities.

Conclusions

This chapter has discussed the origins and impacts of a European network on string theory and gender. The key successes of this network were initiating dialogue about diversity issues within the European string theory community. The network collected the first quantitative and qualitative data on gender which provides baseline data for future studies. The transition from doctoral study to postdoctoral researcher was identified as a key stage at which women researchers are lost even though data show that women are not less successful in postdoctoral applications. It would be interesting to understand in depth why women are more likely to decide to leave the field after their doctorates, and the research discussed in Chapters 4 and 6 is important in this regard. However, more qualitative research on this area in countries across Europe is needed.

The EU-funded COST network concluded in 2017. Following its conclusion, the network's leaders founded Gender in High Energy Theory (GenHET), a permanent working group on high energy theory and gender whose digital presence is hosted by the particle physics laboratory CERN.

GenHET (2022) builds on the work of the COST network: its goals are to monitor the representation of women; to increase knowledge of gender issues and to collaborate with experts on gender to share expertise and good practices, and to provide networking and mentoring for women; and to increase the representation of women at all levels, but particularly in decision-making roles.

The COST network in string theory has had ongoing impacts on the research community. Before the COST network women string theorists were not routinely included as institutional leads for European networks or on conference advisory committees. Following the COST network's inclusion of women, representation of women in roles across the string theory community has improved. For example, it is now expected that at least 10 per cent of speakers in a conference in Europe will be women and increasing numbers of departments monitor the gender balance of their seminar series. Ten per cent is evidently still very low and accordingly many conference organizers set themselves higher targets. In 2022 several major European string theory conferences had over 20 per cent of their speakers being women. However, such a fraction is well above the current fraction of researchers in the community who are women. It would seem challenging to increase this percentage further without backlash from men who feel their chance of being invited to speak is lower than that of comparable women researchers. Analogous issues arise for leadership and advisory roles within the research community: one cannot easily increase the representation of women beyond a ceiling of around 20 per cent without the fraction of women coming through the pipeline increasing substantially.

Diversity issues are increasingly discussed by the string theory community. The main annual conference for the research field, 'Strings', held its first session on gender in string theory during the 2017 Tel Aviv conference. In 2020 the 'Strings' conference included a session on diversity covering race as well as gender. The 2021 'Strings' conference included a panel discussion on gender and diversity, comparing the perspectives of researchers from different generations to explore how the field had progressed. Sessions on diversity are also being included in many other string theory conferences and workshops, showing that the field is increasingly open to discussing these issues. Such sessions are well attended by senior and junior researchers from all around the world. Early career researchers are especially enthusiastic about exploring evidence-based approaches to diversity and inclusion, particularly actions and initiatives to which they can contribute.

Since networking and mentoring are consistently shown to be key to retention and progression of under-represented groups, the community has consciously been expanding its support for mentoring and dedicated networks for minorities. For example, the European string research community has launched a mentoring scheme with senior academics from within the field supporting junior researchers from all over Europe. This scheme incorporates good practices of mentoring schemes run by research councils and universities around Europe, and adapts these to the specific context of string theory that is a field in which obtaining funding and positions is extremely competitive.

Increased engagement with diversity is also evidenced by senior researchers across Europe including the theory department of CERN approaching the leaders of the EU COST project and its successor GenHET for advice on good practice and reports on gender data.

The COST project has thus succeeded both in increasing awareness of gender issues within the field and educating researchers about research on gender inequalities in science. Gender imbalance is now firmly on the agenda of the string theory community. The leaders of the COST project are continuing their work on gender and string theory through a range of projects and initiatives, particularly through the permanent CERN GenHET working group. Funding for meetings that incorporate gender and science has been obtained from research councils and from a number of European research institutes, including Nordita, the Nordic Institute for Theoretical Physics.

Researchers who work on gender and theoretical physics nevertheless face many challenges. The importance of diversity and inclusion work is generally appreciated but, nevertheless, work on gender issues is valued less than physics research outputs in the context of academic appointments and promotions. Research funding for work on gender and science may be viewed as less prestigious than research funding for physics, even though funding for interdisciplinary work on topics such as gender and science may well be scarce and competitive.

It is interesting to note that there is relatively little focus within the string theory community on increasing the diversity of doctoral researchers. String theorists are highly active in outreach, including outreach activities that are targeted at under-represented groups. However, there are rather few initiatives aimed specifically at attracting under-represented groups into doctoral research in string theory, perhaps because doctoral positions are already heavily over-subscribed.

Substantial progress on gender issues in string theory would require more commitment across the research community. The previously discussed mentoring scheme is an example of the type of collaboration required: many men and women senior researchers from institutions across Europe have volunteered to be mentors. In fact, the string theory community has a track record of institutions collaborating for the benefit of the field, such as the research field having agreed annual timelines for offers to postdoctoral researchers. A key aim for the future would be the research field agreeing to common goals and targets on gender diversity and working together to achieve these.

References

Bol, T., de Vaan, M., and van de Rijt, A. (2018) 'The Matthew effect in science funding', *PNAS*, 115(19): 4887–90. DOI: 10.1073/pnas.1719557115

COST Action MP1210 (2022) *The String Theory University*, Available from: www.cost.eu/actions/MP1210/; www.weizmann.ac.il/stringuniverse/

GenHET (2022) *Gender in High Energy Theory*, Available from: https://gen het.web.cern.ch

Gender Net Plus EU (2021) *Report on Gender Equality in Research Funding*, Available from: https://gender-net-plus.eu/wp-content/uploads/2021/ 04/GNP-Deliverable-D6.3-Gender-Equality-in-Research-Funding-plus-Country-reports-final.pdf

Gender Net Plus EU (2022) *Policy Brief – Promoting Gender Equality in Research Funding*, Available from: https://gender-net-plus.eu/wp-content/uploads/ 2022/02/Policy-Brief-Gender-Net-Plus-2022-02-01.pdf

Irish Research Council (2022) *Gender Strategy Review 2022*, Available from: https://research.ie/assets/uploads/2022/03/IRCGenderPlan-s.pdf

Joint Postdoc Applications (2022) *Joint Postdoc Applications Related to Theories on the Unification of Fundamental Interactions*, Available from: https://itf.fys. kuleuven.be/postdoc-application/instructions

Royal Society London (2020) *Diversity Report 2020*, Available from: https:// royalsociety.org/-/media/policy/topics/diversity-in-science/annual-divers ity-data -report-2020.pdf?la=en-GB&hash=34C1EE2FB59D07931A16B 14D806C4A2E

Swedish Research Council (2020) 'A gender equal process – a qualitative investigation of the assessment of research grant applications', Available from: www.vr.se/download/18.6bd0597171d2a04c52e2/158772625282/ A%20gender-equal%20process_VR_2020.pdf

UKRI (2020) *United Kingdom Research and Innovation: Diversity results for UKRI Funding Data 2014–15 to 2019–20*, Available from: www.ukri.org/ wp-content/uploads/2021/03/UKRI-300321-DiversityResultsForUKRI FundingData2014-20.pdf

UKRI (2022) *United Kingdom Research and Innovation: EDI Funding Data*, Available from: https://public.tableau.com/app/profile/uk.research.and. innovation.ukri./viz/EDIfundingdata2021/Awardrate

9

Have Equality Awards Existing in Higher Education in the UK Benefited Current Women Academics?: A Personal Reflection of How My Career Path Has Been Shaped by the Evolving Equality Diversity and Inclusion Agenda

Nicola Wilkin and Jaimie Miller-Friedmann

Introduction

I, Nicola Wilkin, began my undergraduate degree in physics in 1987, unaware that being in a minority as a woman student might influence my career trajectory – or that this issue was of concern to the few senior women academics in the UK. I am now a professor of physics, director of education for engineering and physical sciences at my university, and Chair of the Institute of Physics (IOP), Juno panel (national equality award).

I have advanced through my education and career in an ever-evolving framework of equality schemes that have until recently primarily focused on removing disparity of opportunity for women students.

This chapter reflects on how these schemes have both benefitted me personally and what they have detracted me from doing – and what lessons might be learnt if setting up such schemes from new elsewhere in the globe. This chapter also engages with aspects of the equality agenda that have not been as actively promoted and lobbied for till recently.

The contextual information is authored by Dr Jaimie Miller-Friedmann, my postdoctoral researcher on my current grant 'Centering Women of

Color in STEM: Data-Driven Opportunities for Inclusion' (NSF Grant #1712531/UKRI Grant #ES/T010290/1); my own views do not reflect those of my organization (past or current).

Jaimie has moved back and forth in her career between science and education, teaching physics and biology, and working in laboratories. Her experiences led her to focus on science education and women in science, particularly in physics. Her doctoral thesis was an exploration of the experiences of the most successful physicists and biologists in the UK, uncovering commonalities by field and insights into how and why some women stay in male-dominated fields.

The beginnings

The commentary in this chapter is based on existing within the UK education system and focuses on the evolving opportunities and hurdles within its higher education framework for women in physics. Some of these challenges will have analogues globally, whereas others will be particular to the UK.

The UK educational system is in some ways unique. The system consists of four main parts: primary education (ages 4–5 through 10–11), secondary education (ages 11–12 through 15–16), A-levels (ages 16–17 through 18–19), and higher education (university). During secondary education, students choose to narrow the scope of the courses they are taking, in order to focus on the subjects that will be most important to them in their futures. For example, a student who thinks they're likely to seek out a university degree in physics may choose to study only science and mathematics during their last two years of secondary school. In addition, it is during this stage that students sit the GCSE (general certificate of secondary education) exams, formally known as O-levels (general certificate of secondary education ordinary level exams). Once the GCSE exams are completed, students have a choice to either continue their academic education or to leave this path and either attend further education (often resulting in applied skills useful for hands-on careers such as electrician or carpenter), or apprenticeships, or may decide to work. The GCSE exams simultaneously serve as exit exams (for those who choose not to continue their formal education) and as 'points' indicators for applications to university.

Students who choose to continue their formal education then move on to A-levels, either at the same secondary school, if they are offered there, or at a college (in the UK, a college is where one goes for A-levels or further education, but very rarely higher education). One cannot attend university without completing A-levels. Again, the subjects taken are limited to those most useful to the degree the student intends on earning at university. At the end of the two years, students sit the A-level exams and apply to a course in university. Unlike in other countries, universities in the UK (for the most part) do not have liberal arts programmes. Instead, each student applies to a course (what in the US is a 'major'). Students have been focused on preparing for studying these subjects for four years. In contrast to universities in other countries, in the UK, an undergraduate degree in most subjects is three years. Science students may choose to spend four years at university, and the fourth year is generally spent doing research alongside academics.

The educational system in the UK has gone through vast changes over time, especially within the last 100 years. The current system has been in place for less than half that time, and it is important to note that while thinking about women in academic science, women have only been able to attend and graduate from all UK universities for approximately 50 years.

Women, in a variety of fields, have made great strides in establishing their place in science, technology, engineering, and mathematics (STEM), from inauspicious beginnings effectively banning their participation to women's contemporary status as majority participants in particular fields. In the UK, World Wars I and II changed women's participation in higher education: in the 1920s, 20 per cent of university students were women; by 1962, 9 per cent were women; and by 1992, 37 per cent were women (Hicks and Allen, 1999). Statistics from 2021 show that 57 per cent of undergraduate university students in the UK are women (HESA, n.d.a).

There was, however, a point in time at which physics was populated solely by men. One hundred years ago, women were neither allowed to attend university in the same capacity as men, nor were they allowed to receive degrees confirming the completion of a course load (Historic England, 2024). Inability to obtain a degree in physics made it difficult to pursue an academic career in physics; even if a woman could study the field, she could not have the same honours, titles, or positions as a man. The University of Oxford, for example, did not allow women to graduate before 1920; women who had completed courses prior to that year could return in order to have their degree conferred on them (Oxford, 2013). Once universities opened to women,[1] women *could* pursue careers in academic physics. However, admission into university required a student to have completed their preparatory academic work to that point. The Fisher Act of 1918 had raised the mandatory leaving age to 14, but there were few secondary schools in place, and money was not provided to local authorities to establish secondary schools (Sherington, 1976). In addition, curricula were gender-based: boys received a standard education of fact memorization and instruction in 'manual' skill, while girls received less factual instruction and more lessons on domesticity and needlework (HM Stationery Office, n.d.). The Butler Education Act of 1944 changed education for women from all socio-economic statuses in the UK by raising the mandatory leaving age to

16 for both genders, thus requiring girls to complete preparatory studies for attending university (Middleton, 1972; Armytage, 1978).

In 1944, after many years of advocating for changes in education, the Butler Education Act was legislated. This Act suggested a number of modifications to the education system at that time, including localized authority, an increase in the numbers of schools, a tripartite system of schooling, free milk for children, and a mandatory leaving age of 16 (Middleton, 1972; Armytage, 1978). In addition, although curricula are not explicitly mentioned in the Act, the 'Green Book', a compilation of parliamentary proposals for 'Education After the War', suggested that boys and girls have the same curricula, with gender-associated courses (woodshop or embroidery) added (Armytage, 1978). The Butler Education Act standardized a leaving exam for all students, ensuring that the Green Book proposals gave girls the opportunity to pursue university degrees and academic careers (Middleton, 1972; Armytage, 1978).

From meritocracy to progressivism to the Thatcherism of the 80s[2] (Arnot et al, 1999; David, 2015; Martin, 2022), a revolution in gender slowly began to take shape, allowing women to pursue their chosen careers, not just for 5–10 years before they would start a family, but in a manner equal to men. Laws and economic rewards were put into place to approximate equality in numbers and equity in behaviour (Arnot et al, 1999; Bix, 2004; Kohlstedt, 2004). During the late 60s and the 70s, the UK educational system began to act in terms of meritocracy, a series of reforms that rewarded the individual, and were thought, at least by feminists of the time, to be steps towards equalizing the academic playing field. However, meritocracy served to reinforce the status quo rather than uprooting sexism: Moi gives an assessment of Bourdieu's assassination of meritocracy as smoke and mirrors that make it appear as though coveted positions of power are distributed according to merit (Moi, 1991). In reality, the power in academia remained with men, as they were the only ones with established academic capital; women's achievements were, on the whole, overshadowed by men's (see female scientists from Rosalind Franklin to Jocelyn Bell Burnell to Maria Goeppert-Mayer) and women were not hired as scientists with the same frequency as men. Progressivism, an educational theory that insisted that each child progresses at their own rate, also backfired for girls, since it was in direct conflict with normative ideological femininity. The emphasis placed on girls to be 'good', to be seen and not heard, and to never be aggressive, confrontational, or adversarial, clashed with the need to push against the norm and to succeed beyond expectations, both of which were necessary to study science fields (Harding, 1986; Arnot et al, 1999). While social cultures began to play with ideas of androgyny and transgender lifestyles, by the late 70s science remained steadfastly masculine. Progressivism did help some women to succeed in science, but it did not bring about the equality the UK government hoped it would. By the late 70s and early 80s, social sciences had relatively equal numbers of men and women choosing to major in most fields within social sciences, and although biological science majors were approximately 30 per cent women, physical science majors were 15 per cent women, and engineering majors were about 5 per cent women (Bix, 2004; Thurgood et al, 2006). The 80s were an extremely confusing time in almost every way: a new wave of political conservatism that infiltrated almost all public and private arenas served to simultaneously set ideology back, almost to the 50s and 60s culture of stark binary opposition of the sexes, and implode a new liberalism in music, art, education, and life choices. Thatcher emphasized Victorian values and the nuclear family, effectively eradicating any female aspirations towards a lifelong science career that would serve as the 'centre of one's life' (Arnot et al, 1999).

During this period, feminists in key positions of political power (both men and women) pushed legislation through to enactment so that women had legal rights in the workplace and a fair chance at succeeding in academia. The UK passed the Equal Pay Act in 1970 (Allan, 2011), mandating that equal pay be given for equal work regardless of sex. While this was a step in the right direction, the Equal Pay Act hardly equalized the Bourdieuian 'playing field' of academia. The UK took a decisive step next in 1975 by passing the Sex Discrimination Act (SDA), which prohibited any kind of discrimination (indirect, direct, harassment, victimization – all whether or not the action is perceived or 'real') based on gender (Arnot et al, 1999; Kohlstedt, 2004). This legislations made discrimination illegal and gave women the right to do a great many things, such as study whichever subject they chose, but it did not give them the social, cultural, or ideological capital to do so; women could study all STEM subjects if they wanted to, but only the women who were brave enough to face a social backlash would do so (Harding, 1986).

Again, in the UK, students are imbued with a great deal of responsibility for their own education. The subjects one chooses to study are selected at a relatively young age, and students rarely shift to a different subject once that choice is made. For example, the decision to take a physical science route is taken at 16, as the English system generally expects that the final two years of school are focused on three or four subjects. There are opportunities to take the International Baccalaureate, but this is expensive for schools to deliver, and so the opportunity for a student to keep their options open tends to be restricted to those at private (fee-paying) schools. Unintentionally, this puts a gender bias on the class cohort for physics in the UK. It is different to that for mathematics (popular in itself and as a facilitating subject for economics-related disciplines) and chemistry (required for medicine entry); leading to physics classes at A-level being predominantly male, at an age when students are very aware of the gender dynamics within a class (Smith, 2011a, 2011b; Mujtaba and Reiss, 2013).

Why did I choose to take a physics degree? After all, I had attended an under-resourced comprehensive school,[3] where for instance the final examination chemistry practical was only the second practical we had attempted (Brighouse, 2002). The physics practical had similar challenges – I clearly remember an oscilloscope being shown to us and demoed for the first time before the practical (we were not allowed to touch it, in case we broke it). The physics teachers, who had undertaken physics degrees, were challenged by my questions and need to 'really understand' the concepts beyond the syllabus. Inspiration could be found in our further mathematics teacher – a small class, where he enjoyed any mathematical puzzles we brought to him.

However, I knew I wanted to take physics. Looking back, I had a significant advantage: a physicist father, who had encouraged my interest and curiosity throughout my childhood. Further, there were often PhD students, including women and international, for Christmas festivities at our house. I had no idea of the concept of 'role model', but clearly these were present in a way that was not typical for my peers (Hazari et al, 2007; Archer et al, 2016).

I was persuaded by a friend to join her and her sixth-form college on an outreach visit to Cambridge (my school had no visible ambition in that direction). It was a lovely welcoming day, bar the 'informal chat' with a tutor – whose single utterance to me was that 'he wasn't sure they took students from schools like mine', and in my stubbornness, that was Oxbridge off the list – they didn't deserve me! This incident has always remained with me as a reminder of the disproportionate influence a single conversation can have when one is in an advisory role.

The physics degree I undertook at Southampton was fun and interesting – and I was never conscious of it being detrimental to me that I was a female; my gender didn't affect the behaviour of other students, staff, or demonstrators. From talking to peers retrospectively, I was either very lucky, or just quietly confident in my ability and, alongside that, oblivious to any incidents.

My undergraduate department itself was relatively small and nurturing – I was encouraged to consider a postgraduate degree – and offered good advice on potential supervisors. I chose Manchester, a primary driver being that my main sport was climbing – and this would be a good location! I was much more conscious of being different as a female as a postgraduate and receiving unwanted (low-level harassment) email messages from peers began at that point. As an undergraduate, the field you are studying is a part of your student identity (along with your friendship groups and extra-curricular activities). As a postgraduate, although this is only obvious to me in retrospect, the physics department transitions to a place of work rather than just 'education'. Associated with this, in an era where equality and diversity were not spoken-about concepts, the dynamics of being an isolated (junior) female were just the status quo, and I did not think to question it.

Once I had established myself as a physicist within my cohort of postgraduates, I got on and did physics, and became resilient to the very male environment I was navigating. My extracurricular activities of climbing and caving were also predominantly male – and I found my way through the PhD, with the inevitable highs and lows that come with endeavouring to undertake research. Conferences proved an exception; students and academics from other universities were problematic in their unwanted approaches. However, long before 'active bystander' training my Manchester peers (all male) were protective – and ensured I was never left alone in a vulnerable position.

Most relevant to the equality agenda, I was recruited by Helen Gleeson,[4] then a newly appointed lecturer to participate in an outreach activity for female school students. The programme looked exciting, but I remember a very heated discussion with her, as to why I didn't think it was appropriate to have a single-gender activity, and that I couldn't see why it was necessary or what it was going to achieve. She was very gracious and took the time to discuss and persuade me.

Again, in retrospect this was a very influential moment! I was completely unaware of the discussions that were taking place among female academics as to the challenges and career blocks that appeared to be gender-specific.

Contrast this to my daughters (who are both undertaking STEM degrees, not physics) – who have had frequent input through their time at school both that inequity exists and is recognized – and that there are many programmes and opportunities to proactively ensure that female students' achievements are commensurate with their ability and not detrimentally affected by 'being in the minority'. How much of this is the influence of national initiatives including Athena SWAN and Juno?

The advent of Athena SWAN

By the mid-1990s, female academics in science, technology, engineering, mathematics, and medicine (STEMM) were still not progressing at the same rate as men. Moreover, the gender proportions of undergraduates had changed dramatically, such that women were now approaching, if not meeting, equity, with regard to participation in some fields. These participation levels were not reflected in the most senior positions, and a group of academics decided that, for the sake of social justice, this needed to change. The Athena Project was formed by female academics in STEMM fields, with the intention to 'Advance and promote the careers of women in science, engineering and technology in higher education and research and to achieve a significant increase in the number of women recruited to top posts in the UK' (Athena Forum, 2009; Equality Challenge Unit, 2014; Team et al, n.d.). Initial funding for the project, which was to identify barriers and good practices, came from government-associated bodies (Athena Forum, 2009; Higgins et al, 2014).

The Athena Project consisted of two phases: first, identifying issues and good practices; and second, the development of ways to measure good practices and culture change. Each phase involved several programmes and lasted several years. Phase One ran from 1999 to 2002, and had three programmes, the first of which were the University Development Grant Programmes. The 1999 programme invited bids from universities to research the aims of the Athena Project; six grants were awarded to a varied group of universities, all of which were expected to match funds and to have the approval of their vice chancellor for the project (Higgins et al, 2014). The 2000 programme funded five projects focused on culture change and women's progression and transition from postdoc to faculty (Athena Forum – The Original Athena Project, n.d.; Athena Forum, 2009). The second set of programmes for Phase One involved the establishment of local networks for early career researchers, funding two networks in 1999 and another three in 2000 (Higgins et al, 2014). The last set of programmes was the founding and awarding of the Royal Society Athena Good Practice Awards, which recognized efforts made in addressing gender inequality; these good practices were disseminated in the hopes that they would be beneficial to other universities (Athena Forum, 2009; Higgins et al, 2014).

My husband is also a physicist, and our 'two-body' problem was solved when I secured a lectureship (tenured post) and was able to join him at the University of Birmingham. I saw this as entirely our domestic issue – rather than a recognized career challenge, for academic couples (Harvey, 1998; Schiebinger, 2010; Tzanakou, 2017). My new post was an unusual one, in that it had been created to put in place leadership for student recruitment to the department – in 1999 a physics degree was not a popular choice in the UK. What in retrospect was extraordinary was the leadership I was entrusted with in my first permanent post: create and implement a strategy to improve the tariff (entry grades) of students seeking to join us, and then bring the department with you.

Nowadays, there is significant infrastructure, and leadership is at senior level for admissions processes. In 1999 there was minimal advice, I had had no leadership training other than a funded summer school as a postgraduate, but I had a task, and I treated it (successfully) as if it were a 'physics problem' in itself. This included persuading the head of department that we needed to appoint a schools' outreach lead (the first within the university).

I am now responsible for overall recruitment to engineering and physical sciences. I have been left to wonder if the embedded infrastructure limits agility of response. Even when we think we have empowered more diverse voices, particularly ethnic minorities (in the UK often referred to as BAME: Black, Asian, and minority ethnic) (Writing about Ethnicity – GOV.

UK, n.d.) to be heard, is the ability to effect change stymied by the existence of long-standing systems?

While being responsible for student recruitment I had my first daughter. I did not want to feel like I was abandoning my responsibilities: although no one ever asked what that might mean. On informing my head of department – he was very congratulatory from a human perspective (and as a father himself), but his response was, "well this is the first pregnancy within the department that I've had to deal with –– please talk to human resources and let me know what they advise".

From this point onwards, it is clear how much has changed in two decades – the more likely problem nowadays is that one would need guiding through the wealth of advice available on the internet to ascertain which parts are relevant, and there are visible parenting networks within the university.

Looking back, I struggled with the return to work; it was up to me to choose what I might do in terms of full-time or part-time, but with no case studies, or 'older sister' academics to turn to. Muddling my way through, I wondered if a parent network might not be helpful, and spoke to HR, as to whether I could instigate one, but was rebuffed that this would break confidentiality, and was not the university's responsibility. The lack of parenting support networks was in fact a national issue, and 'Mumsnet', one of the largest parent networks in the UK, was founded at that time (Mumsnet: The UK's Most Popular Website for Parents, n.d.). Nowadays, my university has an active network itself.

With my return to work my sense of risk of letting the side down, and the realization that the playing field was no longer level, crystallized. Becoming a parent is necessarily a life-changing, and (mostly) wonderful, experience. That the employer might be supportive, while crucially not putting the displaced load on colleagues, is a much-improved situation in 2021, thanks to equality, diversity, and inclusion (EDI) voices and actions – but that is not to say things are resolved. It must remain an ongoing and evolving conversation, as witnessed by the home-schooling and work balance that has been negotiated with different levels of success during the UK COVID-19 lockdowns. The impact of this 'everything's at home' year on research and grant-writing and hence career progression is yet to be seen. The research in the general population in the UK showed that the disproportionate load of home-schooling and additional domestic duties fell on women (Kramer, 2020; Staniscuaski et al, 2020; Rosa, 2021).

My second daughter was born in 2003.

Phase Two of the Athena project ran from 2003 to 2007, and consisted of four sets of projects, most of which were concerned with raising awareness of gender inequality in universities as well as rewarding those who were trying to approach equitable systemic solutions. The first set of programmes involved the development and dissemination of the Athena Framework for Action and

Good Practice Checklists in 2003 (Athena Forum – The Original Athena Project, n.d.; Athena Forum, 2009; Higgins et al, 2014). These checklists were distributed to universities, and the data collected showed a lack of awareness of current good practices in place within departments, as well as a lack of commitment to change from heads of department and principle investigators (Athena Forum, 2009; Higgins et al, 2014). The checklists were modified and adapted to become the 'Framework for Action'; this framework is the spine of the Athena SWAN charter, and the IOP Juno Code of Practice (Higgins et al, 2014). The framework identifies five action areas for good practice that emphasize female faculty progression and equitable practice for child and caring responsibilities (Athena Forum, 2009).

The second project was the Athena Survey of Science, Engineering, and Technology, run by the University of Bristol and the University of East Anglia. The survey explored the factors that had previously been identified as important to career progression (Athena Forum, 2009; Higgins et al, 2014). Between 2002 and 2006, the survey was run in a total of 40 universities, and results both confirmed previous factors and identified gender-based bias and assumptions that were possible reasons why female academics did not progress at the same rate as their male counterparts (Higgins et al, 2014). Eventually, in 2006, the survey was made open access, and the project could claim national findings that could be both representative of the UK academic climate, and comparative with regard to best practices (Higgins et al, 2014).

The third project involved collaboration with the Royal Society of Chemistry and the IOP. Both organizations were supportive, financially and in obtaining access to departments. The IOP worked with the Athena Project to develop the Juno Code, which would address the same five pillars of action and good practice, but more specifically within physics departments (Athena Forum, 2009; Higgins et al, 2014; Team et al, n.d.).

The last project, launched in 2005, was the Athena SWAN Charter and Awards Scheme. This Charter emerged from a conference organized by one of the local networks established in Phase One (the SWAN [Scientific Women's Academic Network] Network). The ten founding members of the scheme were selected from the Athena Project's university network (again, established in Phase One). The network members created the charter and the complimentary award scheme, such that Project Juno awards were comparable (Athena Forum – The Original Athena Project, n.d.; ATHENA FORUM, 2009; Higgins et al, 2014; Team et al, n.d.).

Once through the logistics of baby and toddlerhood, our two daughters have always been very close and supportive siblings and this enabled us to travel for physics workshops. We knew they would always be a team in whichever childcare setting I had managed to 'slot them into' on a temporary basis. In theoretical physics, the discipline relies upon extended workshop gatherings of international participants – where a mix of seminars seed informal discussions and collaborations. These workshops are typically a minimum of three weeks and can be up to six months if one is leading a programme. Exemplars of this are Kavali Institute of Theoretical Physics: 'Duration: 2–6 months [KITP strongly discourages visits less than three weeks long (except for experimenters/observers). If you would like to stay less than that, please look for an associated conference]' (KITP website, 2021). Hence, it is imperative for a career in theoretical physics for one to be invited and participate in these workshops. The additional cost of paying childcare for a second time over was, however, a significant issue: the UK childcare costs must be paid to retain a place. Although we could 'manage' it, I felt it was an injustice – after all, no-one expects a researcher not to have their hotel bed paid for, just because they already had a bed at home (Deitch and Sanderson, 1987; Van Anders, 2004; Jons, 2011; Tzanakou, 2017).

This led to my first national lobbying on behalf of fellow early career researchers with caring responsibilities. I was at the time a very active committee member for a number of IOP groups, including Women in Physics, and persuaded the IOP that the survey should come from them; and further they agreed to pay for the analysis of the results. The result was Council agreeing to create and fund the carers' fund.[5]

Looking back, I am surprised at the naïve belief that I could make change happen and the lack of understanding of the depth of bureaucracy I was pitching myself against won out. It should be noted that the childcare issue has also been resolved at the KITP end, with generous provision now in place to support visitors.

Going up against the system was probably not beneficial to me short term as it detracted from research – the main promotion measure within my university at the time. Further, one's reputation among one's research peers within physics is measured via publications and the research grants that facilitate this.

A hard-learnt lesson from academic-equality activity is that the effort and time commitment of the early enthusiasts were not realized and recognized by the institutions that were having 'change done to them' till the schemes had matured.

For schemes being set up elsewhere in the globe, or for more recent university engagement such as the 'Race Equality Charter', the question has to be who is really doing the work and how is the institution going to recognize their efforts; and ensure they are supported when in questioning the status quo they inevitably upset or make their colleagues, and potentially senior management, uncomfortable. Well-nuanced work models for individual academics can be facilitating rather than divisive. These are generally in place in the UK – although what 'best practice' is, is still in development.

Long term, these strategic civic projects were a practice run to where my career has led me – in overseeing the strategic educational agenda for a portfolio of STEM disciplines and being willing to make change happen – even when the bureaucratic hurdles seem impossibly high.

As will become clear, by working to improve equality for the community, I was unintentionally providing myself with CPD opportunities. This is a somewhat overlooked positive of those in minority groups being actively involved in creating change but mentoring alongside it brings even greater rewards to all (Rosser et al, 2019; Ovseiko et al, 2020).

The Project Juno awards

In the midst of the Athena Group's second phase of action, the IOP committed to doing their own research on gender inequality in physics and astronomy departments. The impetus for this action was a lecture given by Jocelyn Bell Burnell, at the Standing Conference of Physics Professors; she described a programme being run by the American Physics Society (APS), entitled 'Climate for Women in Physics Site Visits' (Climate for Women in Physics Site Visits, n.d.; Main et al, 2009). The APS began doing site visits to physics and astronomy departments in 1990 in order to assess the climate of the department, determine how the climate could be improved, and provide assistance to departments in making positive change (Climate for Women in Physics Site Visits, n.d.). The IOP was inspired by Bell Burnell and the APS' well-established programme, and between 2003 and 2005 began a series of research-based site visits (Main et al, 2009). Unlike the APS site visits, which must be initiated by a request from a department chair (for example, please come evaluate our progress and let us know what we could do better), the IOP site visits were offered to all UK physics and astronomy departments as an objective visiting panel that would advise on gender-related issues (Climate for Women in Physics Site Visits, n.d.; Main et al, 2009).

Each institution that agreed to be assessed was visited by a panel that met with senior staff and academics, women academics and students, and a selection of men that were age-matched to the women academics and students (Main et al, 2009). The panel was given a tour of the department and laboratories, and had lunch with female undergraduates (Main et al, 2009). Each visit was followed by a report that highlighted good practices already in place in the department, areas for improvement, and suggestions as to how to implement the actions for improvement (Main et al, 2009). In addition, the departments were required to provide gender-disaggregated data to the panel prior to their arrival; the Juno report notes that many departments had no access to gender-disaggregated data – this was likely the first time the department had considered finding evidence for inequality (Main et al, 2009).

While Project Juno and the Athena Project were similar in intention – identifying good practice and areas for improvement – Project Juno's specificity elicited more personal assessments for departments. Moreover, Project Juno published a report delineating 41 issues common to physics and astronomy departments as well as good practices to eradicate those issues (Main et al, 2009). Project Juno also considered what kinds of issues were specific to women at different stages – undergraduate, postgraduate, early career, and professor. Within the six Principles for Project Juno, equity for all genders is emphasized, the onus for cultivating an inclusive environment lies on the senior management, and it is clear that the vision of Project Juno is to create equity structurally; that is, the institutional commitment to equity is primary, and organization, structure, and management must reflect this (Diversity team, 2020).

Project Juno and the Athena SWAN awards created reciprocity for their awards, such that those departments that received Practitioner or Champion for Project Juno can convert those awards to Bronze or Silver Athena SWAN awards (Diversity team, 2020). The principles on which both awards are based are similar, but Athena SWAN is university-wide, and well known to academics in every field within the UK, whereas Project Juno is more familiar to academics in physics and astronomy, and perhaps other STEMM subjects.

My department was not an initial adopter of either Juno or Athena SWAN. That is to say, there was more interest in the equality agenda than the paperwork required for creating a submission. As it became an embarrassment that departments, we believed, had less interest in equality were being rewarded for their efforts, I agreed to take up the baton of creating relevant working groups and committees and identifying the relevant data. This included lobbying to get university data retrofitted with a 'gender flag'! Previously, although it was self-evident just from looking at a class, that the gender balance was different between disciplines, and it was kept track of at point of admissions, there was no means of tracking attainment by gender. Nowadays, the data sets are automatically far richer in terms of student profiles. Small details such as this genuinely demonstrate the change in perception and importance of monitoring EDI over a decade. Working on the EDI agenda also gave me the chance to work with the person responsible for employment contracts (HR). The insights I gained were incredibly valuable – and I acquired a wise friend. Networking within large institutions is tricky, unless you are senior, and it is your explicit remit – or are engaged in a project that enables you to do this. Although the workload was poorly recognized and is still usually underestimated, the networks I built as a result are among my most cherished and most supportive.

Having been frustrated by the paperwork process, I volunteered and was accepted to become a part of the national Juno panel. It made me aware in detail of the wealth of activity that was taking place across physics departments and associated institutes across the UK and Ireland – and that in the same manner that when one moves from taking exams to writing and marking them is a challenge both in terms of time and ensuring fairness and equality of marking for all. A particular issue is when overall resource (as opposed to that specifically allocated to equality) varies with the financial constraints of different universities, requiring one to measure commitment to EDI activity relative to the size and resources of given departments.

Within my university I continued to support my department's strategy, but realized that I needed to elicit change for the institution as a whole for us to move further forward. I became a member of the overarching university committee – and found myself among like-minded people from across all disciplines and professional services. It was also noticeable that the proactive members were disproportionately female – and the bonding went beyond that of a committee, to the discussion between parents (predominantly mothers) that might have typically happened at the 'school gate'. This included passing down school uniforms for our children.

I was still a relatively junior academic; did any of my efforts begin the institutional change I desired? The importance of working towards an inclusive environment and the benefits for all students and staff and the 'health' of the university overall were not truly recognized. As a scientist, I had felt irritated by the unnecessary hurdles that we had had to overcome in extracting data to write the Juno submission. It was clear that if other departments were to fulfil Athena SWAN criteria, we needed data to be readily available rather than it needing to be hunted down. Hence, I worked with our internal planning people and together we created a university data resource and visualization, which would take the necessary data feeds, and enable departments to drill down and understand their challenges, rather than being exhausted with the equality-agenda effort before they had even established what their actual challenges were. This process has happened in different ways across all universities that are truly engaged with Athena SWAN. And if one were to start afresh – in another country, or for another EDI issue – establishing the data feeds centrally is a clear way of ensuring that maximal time is spent on strategy, rather than data collation, and procrastination in deciding how to present and benchmark.

Alongside my EDI leadership, I had become increasingly engaged with education technology opportunities. In particular, I struggled with the question of whether one could find ways to ensure that confidence in physics and mathematics could be improved in female students, by enabling them to assess their own understanding, without the concern that they were being observed (or perceived to be judged) by another human. The approach I took was implemented across a range of STEM disciplines, and ultimately had me promoted to a professorial post within the University of Birmingham.

The most visible part of my leadership in EDI was as local Chair of the International Conference of Women in Physics in 2017. Physics globally has an overarching professional society membership body – the International Union of Pure and Applied Physics (IUPAP) – whose mission is 'to assist in the worldwide development of physics, to foster international cooperation in physics, and to help in the application of physics toward solving problems of concern to humanity'. It is a complex organization, and Working Group 5 (WG5) of the IUPAP was created in 1999 as a resolution of the Atlanta, GA, General Assembly to survey the present situation and report to the Council and the Liaison Committees, and to suggest means to improve the situation for women in physics (see https://iupap.org/who-we-are/internal-organizat ion/working-groups/wg5-women-in-Physics/).[6]

Hosting the conference is an honour, and bidding to host is undertaken at a national level – I was persuaded by my then university deputy Vice Chancellor, Professor Adam Tickell, and the CEO of the IOP, Paul Hardaker, that Birmingham would be an excellent location and that they were fully supportive, both financially and in terms of administrative support, for me to lead a bid. Having two senior male advocates for the conference, who justified the resource allocation, was a strong statement in itself – and recognized internationally.

The output of the conferences, and the networks new and old revived, can be found at https://aip.scitation.org/toc/apc/2109/1?expanded=2109. Through the contacts of our Pakistani team members, we were also delighted to host a surprise guest speaker – Malala Yousafzai (Wade, 2017).

The current state of affairs

As of 2021, the proportion of female participation in physics and astronomy in university has not changed dramatically from when Project Juno and Athena SWAN began 15 years ago. Women comprise 25 per cent of first-year physics and astronomy undergraduates, 24 per cent of physics and astronomy leavers, 27 per cent of postgraduates, 22 per cent of lecturers, readers, assistant and associate professors, and 11 per cent of physics and astronomy professors (HESA, n.d.b). These numbers show small improvement – women made up 22 per cent of physics and astronomy undergraduates in 2005 – but the numbers in physics have not grown in the same way they have in biology or chemistry, in which female undergraduates are either the majority (biology) or nearing 50 per cent (chemistry). Of the school sciences, physics remains, at least quantitatively, stubbornly male-dominated.

At the faculty level, while there has been improvement in progression and retention of women – in 2004, 4 per cent of physics professors were female, whereas they now comprise 11 per cent of physics professors – attrition remains very high. If women were both staying in physics and progressing comparably to their male counterparts, 25 per cent of physics professors would be female. One *might* therefore assert that the existence of Athena SWAN and Project Juno, and the equality mindset that they have generated among those on promotion and recruitment panels, have made an impact by almost tripling the proportion of women who are making professor, and, in fact, there is some evidence that this is true (Graves et al, 2019). However, there is clearly still a long way to go, and distilling the impact of the schemes, the evolution of ideology, or the increase in numbers of women choosing to study physics is an ongoing discussion (Ovseiko et al, 2017, 2020; Graves et al, 2019; Kalpazidou Schmidt et al, 2020; Xiao et al, 2020).

Both Athena SWAN and Project Juno have spent time and money investing in research and intervention programmes as they have tried to identify and remove obstacles in women's pathways to progression. They have, perhaps, been most effective in having departments and universities create their own action plans, which delineate specific ways in which the department or university plans to approach or meet gender-equality goals within their individualized context. These action plans have not only forced departments and universities to openly discuss issues with equity, but to work together to actively try to improve the diversity of their department, and to be self-reflective about their current and potential practices.

However, there have been inequalities generated in the effort to apply for awards and form action plans. Female faculty are finding that they are encumbered with two additional theoretical and practical concerns. The first is the burden of representation, in which female faculty, as the sole female or one of very few in her department, are standing in as representatives for all women. As such, the work required to submit an application for awards to Athena SWAN or Project Juno falls disproportionately on the few female faculty in the department (Evaluation of Project Juno: Summary of Final Report, 2013; Van den Brink and Stobbe, 2014; Caffrey et al, 2016; Gibson and Dyer, 2017; Ovseiko et al, 2017; Tzanakou, 2019; Tzanakou and Pearce, 2019; Kalpazidou Schmidt et al, 2020; Mousa et al, 2021). The time taken to lead teams in preparing applications and creating action plans can be prohibitive to their career trajectories, since time is taken away from research, publishing, and meeting other requirements for promotion. The second concern is the burden of otherness: by talking about gender inequity, and spearheading panels and committees dedicated to unearthing best (and worst) practices with regard to gender inequity, women are making their differences to their male-dominated departments more apparent. Instead of assimilating and finding a way to cultivate a sense of belonging, which has been proven to be essential to female faculty longevity in physics (Traxler et al, 2016; Blackburn, 2017; Cheryan et al, 2017), female faculty are standing out. Even more, female physicists are frequently interviewed, tracked on social media, and featured on campus posters and billboards, making their femaleness within a male-dominated field – their otherness – altogether more noted.

For some women, these simultaneous burdens may cause them to feel as though they are in a liminal state; they are both representative of their department and different to others in it. As representatives of their department, they submit their applications for Athena SWAN and Project Juno, speak about the department in interviews, and post about their research on social media, and are literally the 'poster girls' for their departments on campus. At the same time, their time is spent differently to their male colleagues, with additional responsibilities, and their fundamental difference, their gender, becomes their primary identity. As such, these physicists are both elite members of their department and do not belong within their department.

Change and the future

While it is not mandatory to join either scheme, the pressures to be morally correct, and the fear of being the villain in this era of striving for social equity, almost force all universities to seek membership and awards. The neoliberal capitalist nature of universities also obligates membership, as equity attracts students to universities (Rosa, 2021).

However, membership does not necessarily mean that gender equity is close to fruition at that institution. And this is where both Athena SWAN and Project Juno face criticism: what has either achieved in the last 15 years? It would be challenging to determine which programmes, schemes, or popular shifts in ideology create the most change in the academy. However, it is safe to say that both Athena SWAN and Project Juno were at the forefront of raising consciousness and opening up the opportunity to talk frankly about inequities in academia. Even more, they created space for women to speak and be heard, and for women to take on leadership roles, which women may not have had prior to Athena SWAN and Project Juno. Before the Athena Project, it was almost taken for granted that the female minority in physics would have to cope with inequities; now, it is unacceptable that any female should have to deal with any sort of discrimination or bias. The open discussions that began to delineate the many ways in which discrimination and bias had been institutionalized also formed action plans to ameliorate physics environments for women, and for all physicists. In truth, each point of an action plan benefits everyone in the department, whether it be suitable and convenient toilets, or zero tolerance for verbal harassment. With Athena SWAN and Project Juno came accountability, and with a Foucauldian sense of policing and transparency, responsibility for ensuring the comfort and safety of oneself and one's peers.

Alongside ideological shifts that favour social justice and a raised awareness of everyday bias, Project Juno and Athena SWAN raised awareness of current unjust practices and ways to slowly make academic environments inclusive. In the quest to apply for awards, the creation of an action plan compels the committee to create a structure or scaffolding of future practice; in essence, both Athena SWAN and Project Juno force individuals to be reflective and vulnerable in evaluating their own behaviours and what they experience. Gender equity has not yet been achieved in physics, but Athena SWAN and Project Juno have opened physics colleagues' eyes to their responsibility for social justice.

I am now Chair of the Juno Panel – and with this responsibility has come a time for reflection. Should it be restricted to gender? Further, although it's not been stated, gender within both Athena SWAN and Juno refers to a dichotomy synonymous with biological sex. Should these schemes be extended to all those who identify as female, or even to all genders? If one asserts that being in a minority is an issue in the workplace, which other minorities should one include in such a scheme? With the end goal always being a maximally inclusive workplace.

Do we ensure that the volunteer panel are appropriately recognized for their significant work that is supporting the physics community rather than merely their institution? I'm working in partnership with the IOP, and the outputs of an external consultancy, PwC, to assess possible future strategies.

However, at its heart, EDI within physics is enabled by the community and for the community. As recognition of the importance of equality work is embedded within institutional activity, there is for me a central question. How do we ensure we embrace and facilitate the activities of the junior academics who are not (yet) part of the infrastructure?

One clue is in valuing the work, both in terms of effort and time diverted from the activities for which an individual is principally employed. Within the UK it is now standard to have a workload model – if one can harness this effectively, and appropriately recompense contributions to equality agenda – for example, with a sabbatical thereafter or reduction in administration elsewhere (for instance, imagine being able to pass on one's exam marking for a large class!)

The question I ask myself is, how would the system, of which I am now a leader, support my younger self in the actions that I undertook two decades ago? Many ideas about gender inequity are now part of mainstream activity but were seen then as time-consuming and awkward in the problems they revealed to the status quo.

We now have frameworks in place across physics in higher education in the UK that recognize and value diversity. This enables those, at all stages in their careers, to be able to question their treatment, and how they are valued – and for those questions when raised to be genuinely considered, discussed, and resolved. Through Athena SWAN and Project Juno, awareness of gender bias in physics has been raised, and awareness has opened up dialogue on how to eradicate other kinds of bias. Alongside this, there is an emphasis on re-integrating, and supporting on return from maternity (and paternity) leave, which is seen as a departmental responsibility rather than expecting that superhero juggling of work–life is the only way to succeed as a woman physicist. There is still much to do – achieving equality is a journey, but the UK higher education physics community has travelled a long way since I started my physics degree in 1987, and I am glad to have contributed a small part to that change.

Notes

[1] Oxbridge were among the last of the UK universities to graduate women.

[2] For more on the policies of meritocracy, progressivism, and Thatcherism, and the ways in which they impacted women and their education, see J. Martin (2022). Throughout these eras, educational policies shifted dramatically, making it so that, for example, school attendance until the age of 16 was mandatory for all; before this time, the proportion of women attending and completing their school-age education was far below the proportion of men.

[3] The predominant form of state-funded secondary education in the 1980s in England.

[4] 'I attended Holy Family Comprehensive School in Keighley as a teenager. I then moved to Manchester and graduated with a 1st class Joint Honours degree in maths and physics from Manchester University in 1983. I undertook an industrially sponsored experimental PhD in the optics of liquid crystals having been inspired by an undergraduate lecture course, gaining my PhD in 1986. I spent three years in a rather unusual postdoc role, running an industrially funded research unit at Manchester, an experience that shaped my academic career during which I've continued to interact with industry. I joined the academic staff in physics at Manchester in 1989 as their first female lecturer, and subsequently held a number of posts in the university, including Associate Dean for Research in the Faculty of Engineering and Physical Sciences (2002–07) and Head of the School of Physics and Astronomy (2008–10). In 2015, I moved to Leeds as Cavendish Professor and Head of the Soft Matter Physics Group and was Head of School from 2016 to 2021' (H. Gleeson, 2021).

[5] www.iop.org/about/support-grants/carers-fund#gref

[6] The International Conference on Women in Physics (ICWIP) is held every three years, bringing together men and women from around the world to report on the situation of women in physics in their country, to share good practice, to suggest and implement means of improvement, and to network. Over 92 different countries and over 1,300 delegates have attended the ICWIPs, and many new national bodies on women in physics have been created and regional meetings have taken place. The Conference Proceedings, available online, are a source of statistics and good practice across the world.

References

Allan, E.J. (2011) *Women's Status in Higher Education: Equity Matters*, San Francisco, CA: Jossey-Bass/Wiley.

Archer, L., Moote, J., Francis, B., DeWitt, J., and Yeomans, L. (2016) 'The "exceptional" physics girl: a sociological analysis of multimethod data from young women aged 10–16 to explore gendered patterns of post-16 participation', *American Educational Research Journal*. DOI: 10.3102/0002831216678379

Armytage, W.H.G. (1978) 'Back-up to Butler: the biosocial background of the Butler Act', *Westminster Studies in Education*, 1(1): 5–21. DOI: 10.1080/0140672780010102

Arnot, M., David, M., and Weiner, G. (1999) *Closing the Gender Gap: Post War Education and Social Change*, Polity Press.

Athena Forum (2009) 'A guide to good practice for professional and learned societies 2', Available from: www.athenaforum.org.uk

Bix, A.S. (2004) 'From "engineeresses" to "girl engineers" to "good engineers": a history of women's U.S. engineering education', *NWSA Journal*, 16(1): 27–49. DOI: 10.1353/nwsa.2004.0028

Blackburn, H. (2017) 'The status of women in STEM in higher education: a review of the literature 2007–2017', *Science and Technology Libraries*, 36(3). DOI: 10.1080/0194262X.2017.1371658

Caffrey, L., Wyatt, D., Fudge, N., Mattingley, H., Williamson, C., and McKevitt, C. (2016) 'Gender equity programmes in academic medicine: a realist evaluation approach to Athena SWAN processes', *BMJ Open*, 6(9): e012090. DOI: 10.1136/BMJOPEN-2016-012090

Cheryan, S., Ziegler, S.A., Montoya, A.K., and Jiang, L. (2017) 'Why are some STEM fields more gender balanced than others?', *Psychological Bulletin*, 143(1): 1–35. DOI: 10.1037/bul0000052

David, M. (2015) 'Women and gender equality in higher education?', *Education Sciences*, 5(1): 10–25. DOI: 10.3390/educsci5010010

Deitch, C.H. and Sanderson, S. W. (1987) 'Geographic constraints on married women's careers', *Work and Occupations*, 14(4): 616–34. DOI: 10.1177/0730888487014004007

Diversity team, IOP (2020) 'Code of practice for Project Juno: what the awards are and how to achieve them', Available from: https://www.iop.org/sites/default/files/2021-02/Juno-Code-of-Practice-May-2020-v3.pdf

Equality Challenge Unit. (2014) Athena Swan Charter.

Gibson, V. and Dyer, J. (2017, July) 'Project Juno: advancing gender equality in physics careers in higher education in the United Kingdom', The European Physical Society Conference on High Energy Physics, Available from: https://pos.sissa.it/

GOV.UK (n.d.) 'Writing about ethnicity', *GOV.UK*, [online] 6 January 2022, Available from: www.ethnicity-facts-figures.service.gov.uk/style-guide/writing-about-ethnicity

Graves, A., Rowell, A., and Hunsicker, E. (2019) *An Impact Evaluation of the Athena SWAN Charter*, April, pp 1–171. Available from: www.ecu.ac.uk/wp-content/uploads/2019/08/Athena-SWAN-Impact-Evaluation-2019.pdf

Harding, S.G. (1986) *The Science Question in Feminism*, New York: Cornell University Press.

Hazari, Z., Tai, R.H., and Sadler, P.M. (2007) 'Introductory university physics performance: the influence of high school physics preparation', *Gender and Education*, 91(6): 847–76. DOI: 10.1002/sce

HESA (n.d.a) 'Data collection', Available from: https://www.hesa.ac.uk/collection

HESA (n.d.b) 'Heidi Plus: higher education business intelligence', Available from: https://www.hesa.ac.uk/services/heidi-plus

Hicks, J. and Allen, G. (1999) 'A century of change: trends in UK statistics since 1900', House Of Commons Library, Available from: https://commonslibrary.parliament.uk/research-briefings/rp99-111/

Higgins, J., Lane, N., Fox, C., Leyser, O., Higgins, J., and Bell Burnell, J. (2014) 'The Athena Project Review summary'.

Historic England (2024) 'The pursuit of knowledge', Available from: https://historicengland.org.uk/research/inclusive-heritage/womens-history/visible-in-stone/university/

HM Stationery Office (n.d.) 'The Dyke Report (1906): report of the Consultative Committee upon questions affecting higher elementary schools', Available from: https://education-uk.org/documents/dyke1906/dyke06.html

Jons, H. (2011) 'Transnational academic mobility and gender', *Globalisation, Societies and Education*, 9(2): 183–209.

Kalpazidou Schmidt, E., Ovseiko, P.V., Henderson, L.R., and Kiparoglou, V. (2020) 'Understanding the Athena SWAN award scheme for gender equality as a complex social intervention in a complex system: analysis of Silver award action plans in a comparative European perspective', *Health Research Policy and Systems*, 18(1). DOI: 10.1186/S12961-020-0527-X

Kohlstedt, S.G. (2004) 'Sustaining gains: reflections on women in science and technology in 20th-century United States', *NWSA Journal*, 16(1): 1–26. DOI: 10.2979/nws.2004.16.1.1

Main, P., Dyer, J., Ahmed, S., and Hollinshead, K. (2009) 'The Juno code of practice: advancing women's careers in higher education', *International Journal of Gender, Science and Technology*, 1(1), Available from: http://genderandset.open.ac.uk/index.php/genderandset/article/view/50

Martin, J. (2022) *Gender and Education in England since 1770: A Social and Cultural History*, Palgrave Macmillan.

Middleton, N. (1972) 'Lord Butler and the education act of 1944', *British Journal of Educational Studies*, 20(2): 178–91. DOI: 10.1080/00071005.1972.9973344

Moi, T. (1991) 'Appropriating Bourdieu: feminist theory and Pierre Bourdieu's sociology of culture', *New Literary History*, 22(4): 1017. DOI: 10.2307/469077

Mousa, M., Boyle, J., Skouteris, H., Mullins, A.K., Currie, G., Riach, K., and Teede, H.J. (2021) 'Advancing women in healthcare leadership: a systematic review and meta-synthesis of multi-sector evidence on organisational interventions', *EClinicalMedicine*, 39. DOI: 10.1016/J.ECLINM.2021.101084

Mujtaba, T. and Reiss, M.J. (2013) 'What sort of girl wants to study physics after the age of 16? Findings from a large-scale UK survey', *International Journal of Science Education*, 35(17): 2979–98. DOI: 10.1080/09500693.2012.681076

Ovseiko, P.V., Chapple, A., Edmunds, L.D., and Ziebland, S. (2017) 'Advancing gender equality through the Athena SWAN Charter for Women in Science: an exploratory study of women's and men's perceptions', *Health Research Policy and Systems*, 15(1). DOI: 10.1186/S12961-017-0177-9

Ovseiko, P.V., Taylor, M., Gilligan, R.E., Birks, J., Elhussein, L., Rogers, M., Tesanovic, S., Hernandez, J., Wells, G., Greenhalgh, T., and Buchan, A.M. (2020) 'Effect of Athena SWAN funding incentives on women's research leadership', *BMJ*, 371. DOI: 10.1136/BMJ.M3975

Rosa, R. (2021) 'The trouble with "work–life balance" in neoliberal academia: a systematic and critical review', *Journal of Gender Studies*. DOI: 10.1080/09589236.2021.1933926

Sherington, G.E. (1976) 'The 1918 education act: origins, aims and development', *British Journal of Educational Studies*, 24(1): 66–85. DOI: 10.1080/00071005.1976.9973457

Smith, E. (2011a) 'Staying in the science stream: patterns of participation in A-level science subjects in the UK', *Educational Studies*, 37(1): 59–71. DOI: 10.1080/03055691003729161

Smith, E. (2011b) 'Women into science and engineering? Gendered participation in higher education STEM subjects', *British Educational Research Journal*, 37(6): 993–1014.

Thurgood, L., Golladay, M.J., and Hill, S. T. (2006) *U.S. Doctorates in the 20th Century*, Alexandria, VA: National Science Foundation.

Traxler, A.L., Cid, X.C., Blue, J., and Barthelemy, R. (2016) 'Enriching gender in physics education research: a binary past and a complex future', *Physical Review Physics Education Research*, 12(2). DOI: 10.1103/PhysRevPhysEducRes.12.020114

Tzanakou, C. (2019) 'Unintended consequences of gender-equality plans', *Nature*, 570(7761): 277. DOI: 10.1038/D41586-019-01904-1

Tzanakou, C. and Pearce, R. (2019) 'Moderate feminism within or against the neoliberal university? The example of Athena SWAN', *Gender, Work & Organization*, 26(8): 1191–211. DOI: 10.1111/GWAO.12336

Van Anders, S.M. (2004) 'Why the academic pipeline leaks: fewer men than women perceive barriers to becoming professors', *Sex Roles*, 51(9–10): 511–21. DOI: 10.1007/s11199-004-5461-9

Van den Brink, M. and Stobbe, L. (2014) 'The support paradox: overcoming dilemmas in gender equality programs', *Scandinavian Journal of Management*, 30(2): 163–74. DOI: 10.1016/J.SCAMAN.2013.07.001

Xiao, Y., Pinkney, E., Au, T.K.F., and Yip, P.S.F. (2020) 'Athena SWAN and gender diversity: a UK-based retrospective cohort study', *BMJ Open*, 10(2). DOI: 10.1136/BMJOPEN-2019-032915

PART IV

Personal Experiences

Introduction

Parts I–III of our collection have been devoted to exploring contemporary research and theory which discuss and analyse some of the structural, organizational, and personal mechanisms that limit or reduce the chances of women integrating successfully into the field of physics. The approach throughout has been underpinned by a feminist orientation to the issues raised and to argue for greater improvement in gender equity and inclusion. Key to this commitment therefore is that this book also includes the views and positions of physicists *themselves*, to enable their agency within their representation. In Part IV, therefore, we now turn to the subjective experiences of physicists through a combination of biographical interviews and personal photographs.

In the first chapter of Part IV, 'Biographical Accounts of Physicists', we present a series of personal stories of women physicists, and one man, one of our editors, Yosef Nir, who has worked extensively to promote gender equality in physics. In a series of biographical interviews conducted by Marika Taylor, one of the editors and a physicist herself, Marika reveals the personal and unique accounts of a small sample of people who chose a career in physics. The interviews reveal interesting diversities in their senses of identity, personal choices, and views about gender and inequality.

The second chapter, 'A Room of One's Own: Photographs of Women Physicists in their Working Spaces' edited by Meytal Eran Jona, is dedicated to visualizing gender and physics. It includes photographs of women physicists in their chosen setting, be this their workspace, office laboratory, or within a field experiment. The chapter is called 'A Room of One's Own' in tribute to Virginia's Woolf's iconic book, first published in 1929.[1] Woolf was perhaps among the first women to conceptualize the resources that women need in order to become a successful writer. In her essay, she uses metaphors to explore social injustices and women's lack of free expression – her most essential point being summed up in the recognition that 'A woman must have money and a room of her own if she is to write fiction' (Woolf, 1929: 5).

The diversity of women doing physics is reflected in both the biographical and photographic chapters. Through these chapters we learn that there is no 'one' physics, just as there is no single category of women physicists. There are endless shades of physics, intersecting with the many ways in which physics can be done. Each woman can choose her own path, her own physics, and her own room.

Note

[1]   V. Woolf (1935) [1929] *A Room of One's Own*, London: Hogarth Press.

10

Biographical Accounts
of Physicists

Marika Taylor

The aim of this chapter is to explore personal stories of physicists including their own experiences and views on under-representation of women in physics. Discussions around the under-representation in physics often focus on structural factors that influence the attraction and retention of minorities into the field. In this chapter we focus on individual perceptions of the culture and environment in physics and the common themes that emerge from physicists' experiences.

Four physicists were asked to participate in interviews; they were chosen to reflect different career stages, different European countries, as well different research fields within physics. Following the principles of semi-structured interview, we used an interview-structure guide to ensure organization and flexibility of the conversations. The interview guide included a set of proposed questions. Although these questions were prepared beforehand to guide the conversation, the interview prioritized open-ended questions and encouraged two-way communication. All the interviews were recorded via Zoom and the recordings were used to produce written accounts.

Key themes explored during the interviews included: (i) what attracts and retains researchers in physics; (ii) perceptions of the culture and environment of the physics community; (iii) differences in opportunities for under-represented groups; and (iv) suggestions for actions that could be taken to improve diversity and inclusion.

Yasmine Amhis

PhD 2009 Orsay, France.
CNRS researcher at IJCLab, Orsay, France.
Physics coordinator for LHCb experiment, CERN.

What is your field of physics research?

I work in experimental high-energy physics. During my master's studies I had the chance to go to the international particle physics lab, CERN, as a summer student and this was a turning point in my career. Now I am the lead physics coordinator for LHCb, one of the major experiments at CERN.

Why did you choose to become a physicist?

I originally signed up to medical studies in France; both my parents are medical doctors in Algeria. I did not enjoy the set-up and the programme but the physics within the programme was very good and my interest in physics started from there. The path to move to physics was not straightforward: I had to restart my studies. My parents were initially worried about me changing path but soon realized that I was very happy doing physics.

What do you enjoy about physics?

I enjoy its logic and rigour. I like to look at a problem and break it down into parts that are understood, and parts that are not. As an experimentalist I have experienced the joys of discovery, of being the first person to see data showing a new result.

Work–life balance

My parents have always been supportive of me and are proud of my achievements. My husband and I are both physicists, within the same research field. We both have significant leadership responsibilities, and we ensure that we balance the needs of our careers with looking after our child. For example, when we attend the same scientific conferences we split our attendance days equally when we have our daughter with us.

Mentoring and leadership

Throughout my career I have learned a lot from other people, particularly from reflecting on why I admired how they reasoned or how they presented their work. When I work with students and postdoctoral researchers I help them to identify their goals and to reflect on what would they be proud to achieve. In many ways my approach is very similar to that described by the conductor Benjamin Zander in his books. I enjoy working collaboratively with my team, discovering new paths of research, and seeing them develop. I try to give people the type of support from which I benefited. I was encouraged to apply for leadership roles in experimental teams, roles for which I would not have had the confidence to apply without that push.

Barriers to success

Barriers and obstacles can be subtle. As a student I could see differences between me, whose parents were doctors in another country, and students who had family background in academic science in France and knew the system. As a researcher I notice comments and behaviours that act as obstacles, including by depleting your energy. For example, when I had a child, there were suggestions that I might spend more time with her rather than take up leadership roles with additional responsibilities. There are times when a male colleague's suggestions are received more seriously. In high-energy experiments women do have major roles but it is still not common for them to make it all the way to be the project leader, the spokesperson for a large experiment, and so on

Self-confidence

My confidence as a leader depends on the time of the day. I have moments when I am confident, others when I am not. I have learned with time to build confidence on past experiences and achievements. At some point in my career I almost forced myself to overcome shyness and ask questions in meetings. It really felt uncomfortable the first few times but it paid off.

Attraction and retention of under-represented groups

I am fortunate to have been brought up to believe that there was nothing I could not achieve if I worked hard enough. When I experienced a teacher making a gendered comment of some kind, I had the confidence to see that this was inaccurate and inappropriate. This upbringing helped me overcome any obstacles. It was when I became a mother myself that the gender question really hit me in the face. I can see that there are gender differences even in how young children are treated: my 4-year-old daughter said recently that caps are for boys (nor girls) and she must have learned this from the environment outside our home. At home we try to undo intrinsic biases on a daily basis, reassuring her that she is good at numbers, that she can achieve whatever she wants in life.

There have been some changes in attitudes since I entered the field. If I look at young researchers they are much more aware of the gender question and they are less accepting of casual misogynistic behaviours and comments. Some of the people who were most supportive of me are evolving further on their own journeys, and are supporting the research community in becoming more mindful about how it behaves. On the other hand, most of those involved with diversity initiatives are from under-represented groups themselves or are established supporters of under-represented groups. There are women in senior roles but the numbers are small and the rate of change is slow.

Actions for the research community

It is not easy to identify actions that will be effective at attracting and retaining under-represented groups. However, there are many small things that we can do to improve the culture and the community. We can encourage women to contribute comments and questions, give value to their thoughts, and lift their confidence by positive feedback loops. We can pay attention to scheduling of events and be mindful of the language we use in meetings. Replacing gendered language such as 'manpower' may seem a small thing in itself but this is important in making everyone feel included. In writing reference letters for women, referees need to emphasize scientific and leadership skills, rather than commenting on pastoral and personal attributes. There should be more physics textbooks written by women. Culture change is a collect of many such small things, all done consistently.

Yosef Nir

PhD 1988 Weizmann Institute, Israel.
Professor of Physics, Weizmann Institute, Israel.

What is your field of physics research?
I work in particle physics, the field of physics whose aim is to formulate the basic laws of Nature. I work on theory, interacting closely with experimentalists.

Why did you choose to become a physicist?
I often say that I became a physicist because of my high school teacher, Mario Livio, who became a famous astrophysicist and author of popular science books. In retrospect I feel I wanted to be a scientist and I looked for a scientific discipline. Mathematics was for me too much removed from reality while physics was more interesting than sciences such as chemistry. Neither of my parents worked in science or universities but they supported

me to study. Being a professor is highly appreciated in my wider family and in Israeli society. Physics is a respected discipline; one that is considered difficult and demanding.

When I was deciding on future paths after military service I considered psychology. Careers advisors even recommended psychology as they felt I would have more chances for human interactions. I found out later that including human interaction in your work is a choice, one which is a very enjoyable part of physics for me.

Career progression

After I finished my military service I studied for bachelors and master's degrees in physics. At the end of my masters, I reflected on what to do in a doctorate. I had several possible pathways but ended up working in particle phenomenology. My doctoral advisor had many similarities to myself, both in strengths and in weaknesses. I completed postdoctoral research in a leading US lab, SLAC, before taking up a permanent post at Weizmann Institute in 1990.

Work–life balance

I have a spouse and children. I have never felt a conflict with work and my spouse has always had a career, including her own business. I think she would make similar statements with respect to us balancing our careers. At one point in our life together we said that equality is not necessarily something that has strict criteria every day but is rather something that should be viewed over longer timescales. There were times when I took more responsibility at home and times when my career demanded more and it was the opposite way round. We never hesitated to get help at home, particularly when the children were young. My children would sometimes come to summer research schools with me, and they also came when I went on sabbatical to Princeton.

Achievements, mentoring, and leadership

The achievement of which I am the most proud is that seven of my former PhD students have academic positions in very good places. Most of them go well beyond what I taught them and beyond the field in which they worked with me. I am also often asked to lecture at advanced physics schools around the world; these are key to young researchers' development. Years later established academics will tell me that they attended my lectures as a doctoral or postdoctoral researcher and that I really helped them to get onto their scientific trajectories. In fact this process of teaching in advanced schools has helped me to formulate concepts clearly.

Barriers to success

I cannot recall a period where I have personally felt disadvantaged. I did however start noticing gender differences in the research community from when I was a postdoctoral researcher. One of the physicists in my research group at that time was a highly distinguished woman researcher but she did not have a full professorship. She only became a full professor much later, at the age of 60 or so. In the same period I was working with a young European woman and I still remember comments being made over the lunch table about her winning a physics prize because of wearing a short skirt. At the time I did not how to react; now it would be completely different.

When I took up my post in Weizmann in 1990 I realized that there had never been a single woman permanent staff member in physics at Weizmann. This looked strange to me; there were not many women in the US lab but there were a few. When I started asking people about it, they would say that there was no problem with bias. At the time I felt that the faculty did have a problem but I only started to become more active on this issue later. It was around 2008 when I started to speak up and change the dynamics about appointing women. When I became the Dean of Physics this was an opportunity for me to act in a more serious way. As Dean I learned a lot from talking to women staff and students, from hearing personal stories and small incidents. As a chairperson of the institute-level appointment committee I understood much more about the dynamics of such committees and how such dynamics can influence decisions that we consider to be completely objective. In recent years I started working with a sociologist to understand gender issues in physics.

I do think things have changed. These days a woman with the reputation of my female colleague above would definitely get a professorship much earlier in her career. People understand that they cannot make jokes about a woman's clothes and they are more careful in what they are saying. For some people this may just be recognition of what is permitted to be said but things have changed in terms of both processes and culture.

Self-confidence

I have good friends and collaborators who have great breakthroughs on their records. I think I have very good works but perhaps not breakthroughs. Maybe my potential as a physicist has not been realized to its full capacity. However, I am happy both as a physicist and as a person, as well with the work I have been doing in collaboration with a sociologist in recent years.

Attraction and retention of under-represented groups

The physics community can be very competitive, even aggressive. I feel uncomfortable about that. I am not under the illusion that I am not personally competitive, but my competitiveness is not to better than other people. I want to recognize what I have contributed to the field and then my competitiveness is somehow satisfied.

I want my faculty, my institute, to be as good as possible. I believe that diversity is good for physics: diversity leads to different lines of thought, new collaborations, and new perspectives. In hiring women as well as men, we double the size of the talent pool. In academia we often say that we need to change academic leaders, such as the deans and presidents of institutes, because every leader has strengths and weaknesses. I think different styles of leadership in senior positions and committees is a good thing, and diversity among staff helps realize this.

Actions for the research community

I really believe in one plus one plus one, at least in my environment. This is certainly a very concrete and explicit approach – to take one action at a time to improve gender equality. It might be facilitating one woman to continue in the field after taking a break with children. This might not immediately be a tenure track post, but it could be the first step towards one. One plus one is about pushing things a little bit forward with every action you take.

Hannah Price

PhD 2013 University of Cambridge, UK.
Royal Society University Research Fellow at University of Birmingham, UK.

What is your field of physics research?

I work in theoretical physics, specifically in the field of ultra-cold atoms.

Why did you choose to become a physicist?

I came from a family that was always very much about learning; my parents were both teachers. When I was 9 or 10 a documentary on Fermat's Last Theorem made a huge impression on me. I was awestruck at somebody spending years of their life solving a mathematical problem but I was also taken aback by the isolation of this work. I knew that I wanted to do something like mathematics but that I would also want to work with other people. When I started reading popular science I got really interested in areas such as particle physics and I went to study natural sciences at Cambridge. As I learned more physics I realized that I was less interested in particle physics, with its focus on very large experiments and big data, and more interested in smaller physics collaborations. It took me a while to decide between experiment and theory: the kind of atmosphere around theoretical physics put me off. Theoretical physics was considered a hard area of research and it was only after an internship that I felt confident enough to continue into this area.

Career progression

I was lucky to have a doctoral supervisor who taught me a great deal in terms of how to approach a physics problem, frame questions, and solve them. Yet at the end of my doctorate I was not sure whether I wanted to continue in academia as I found research quite socially isolating. With my supervisor's support I found an ideal postdoctoral research post in an Italian group that was highly collaborative and connected, giving me exactly the dynamic I wanted. After four great years in Italy I started looking for a permanent role which would fit with my partner's career. We were very lucky to find posts for both of us in Birmingham. He is based in material sciences and I have a long-term fellowship funded by the UK Royal Society.

Work–life balance

My department has been very supportive with respect to parental leave. For example, my colleagues have made sure that I do not need to worry about my PhD students or anything like that during leave. My husband and I have taken shared parental leave: six months each. This shared approach to parenting is something that is important for developing better research culture and environment.

Mentoring and leadership

I have had formal mentors through the university and the Royal Society but informal mentoring relationships have actually been more important. I am lucky to have senior colleagues who constantly provide little bits of advice, reassurance, and encouragement. I am coming to a career stage where I will be taking on more leadership roles myself, reaching out beyond my own group. There is still much to do in terms of improving diversity in all senses. For example, it is still the case that most of our doctoral researchers are White, male, and from the region. Increasing our diversity will enrich everyone's experience.

Barriers to success

I have quite often felt a sense of isolation and this has been exacerbated by being one of the very few women in the field. Being one of the few women makes me stand out. As a PhD student I felt it was very hard to socialize with the research group. By contrast my male partner had a different experience, going out regularly with senior colleagues for dinners and receiving informal careers advice.

I had not anticipated some of the challenges of an academic career. When I started university I did not know about academic career paths and the postdoctoral period of insecurity and mobility came as a surprise to me. This is a structurally difficult period where you are expected to have short-term contracts in different countries and this derails a lot of people from pursuing an academic career in physics.

Self-confidence

There is very much a cult of genius around physics. I do not identify with that and this cultural aspect has led me to question whether physics research is the right place for me. Nowadays gender equality is acknowledged as an issue and women are offered more opportunities than they would have been in the past, such as conference talks. However, there can be a perception that you have been offered an opportunity just because you are a woman and this can get inside your head, undermining your confidence.

Attraction and retention of under-represented groups

I studied at a girls' school which was helpful in that gender was not an issue in the classroom. It did, however, mean that I was less used to male behaviour when I went to university, particularly the jockeying for position and competitive laziness. Some men would deliberately give the impression of doing no work but would actually work hard aiming for high performance.

There are cultural differences with respect to women and science. In the UK it is not uncommon to hear people say that girls dislike maths and are not good at it. This is not the same in Italy, where around half of the students on university maths and physics courses are women. Italy has cultural barriers to women continuing in science but a different attitude to women studying science throughout school and university.

I think it is important to realize that not only is physics pushing women away but also other subjects attract them. In school I was very much interested in other subjects. I decided to study physics after finding out more about it and realizing that it was not an area I could explore in my spare time. The fact that physics seems less people-oriented and less directly socially applicable is a factor in attracting women.

Actions for the research community

There are actions that the physics community itself could implement to retain under-represented groups. Summer internship projects for students can lead to people staying in science and feeling themselves becoming part of the science community. Other activities such as summer schools and conferences for students can encourage women to continue into doctoral research and beyond. The actual environment of a physics department is important. Spaces for students to study and socialize can strengthen relationships with researchers in the department and facilitate a sense of belonging.

The culture of postdoctoral positions with short-term contracts and requirement of mobility between countries is difficult. Yet shortening the postdoctoral period could have unintended consequences, with offers of permanent academic posts relying too much on the prestige of your supervisor or university. Fellowship schemes such as the Royal Society Dorothy Hodgkin are designed for those with family needs for long-term funding in one place and are important. However, such fellowships cannot address the wider challenges of balancing family life and dual careers with mobility requirements and high competition for permanent posts.

Petra Rudolf

PhD 1995, Namur, Belgium.
Professor of Physics, University of Groningen, Holland.

What is your field of physics research?
I am an experimental solid-state physicist.

Why did you choose to become a physicist?
I was always curious about Nature. I recall a teacher in primary school explaining glaciation with a sandbox and water. This is when I realized that you could model Nature. When I finished high school I was not yet sure that I would go into physics but my classmates tell me they always knew I would do physics. They remembered me explaining particle wave duality at the blackboard when the rest of the class could not follow the concepts. I got into solid-state physics by chance. In my second year of study I had a boyfriend who was doing a doctorate in experimental physics and I realized I wanted to work in such a lab. Physics research is a field where there are

surprises all the time. Even after working in physics for 20 or 30 years, you are always solving new problems.

Career progression

I come from a family where people go to university. I realize that has been a privilege. My family has also worked in many different countries and this helped because what I was doing was not strange in the family context. It took my quite a long time to finish my degrees as I had to work and study at the same time. I then worked in the US Argonne and Bell labs before returning to Italy. However, it was clear that my contract in Italy would not be continued so I had to look for other positions. I wanted to go to Cambridge but my husband advised me to go to Namur in Belgium. He thought I would have a strong chance of a permanent post there and he was right. The Namur group was small and accordingly the research community knew that I was leading my own work and I got a lot of visibility.

Achievements, mentoring, and leadership

I am proud of the people I have educated and the careers they have made. This has had a much bigger impact than any single paper of mine. There is nothing of which I am prouder than attending the examination of one of my doctoral students. You see that your student who barely knew how to draw a graph a few years earlier is now a master of their field and ready to discuss with eminent scientists. I also love to teach first year students straight out of high school, introducing them to new areas of physics such as nanoscience. I enjoy mentoring young scientists, women in particular, helping them to improve their grant proposals and coaching them for interviews. I am almost happier for their successes than I am for my own.

I use a team approach to science: half of my projects are with my doctoral students. I like to work in different fields and study different problems, collaborating with people from different backgrounds. At times this is criticized as some senior colleagues feel one should specialize in a single field. Specialists can certainly make very important contributions but sharing your professional experience with other fields should be accepted as equally valuable.

Barriers to success

When I started in my career I was not prepared for others being envious. I never thought that scientists would resent success and behave unpleasantly because of my success. When it happens, and it does happen, it gives me the biggest blow to realize somebody can behave this way.

I studied in Italy where undergraduate physics is half male and half female. In Italy many of the women physics students became teachers so there is a problem there as well but it comes later. The first time I realized physics was a male profession was when I did an internship in the US and realized there were hardly any women around.

At the beginning of my career there were comments that probably would not be made today. I complained about my salary being low relative to the high cost of living in an expensive city and I was told that this should not be a problem as I was married. Nowadays there is awareness of the gender pay gap but still women university professors in Holland are paid €400 per month less than their male colleagues. I negotiated well for my current role but I see that many colleagues do not, and they struggle to catch up from their lower starting salary. I give a lot of talks to young researchers about preparing for the next steps of their careers because I think we have a duty to guide them, particularly those from under-represented groups who may not get the same level of mentoring from their colleagues.

Attraction and retention of under-represented groups

Looking at the pipeline the transition from postdoctoral researcher to permanent posts is a particular problem. Sometimes you see people who have done excellent science but have not given invited talks, obtained teaching experience, or applied for funding. They are less competitive for permanent positions but they did not know they had to do these things. This can particularly affect women as they may not get as much advice or mentoring.

When I came to Groningen almost 20 years ago I was the first woman in physics. Now we are 20 across physics and astronomy and you can feel that the atmosphere is different. We have certain things embedded in our processes, such as all members of a hiring committee being responsible for attracting diverse candidates. Such rules oblige everyone to be proactive.

Actions for the research community

I think we need to do cultural training at all levels if we want to be more inclusive as a community. Many group leaders do not know how to deal with difficult issues such as undesirable behaviour or sexual harassment. People are not confident about having difficult conversations. Leaders who are women, or from different cultural backgrounds, often have different management styles. Staff who have never been managed by a woman can feel uncomfortable and ill at ease; training on cultural issues and different leadership styles may help.

Under-represented groups may not understand what they need to do to get promoted. Sometimes expectations will be left implicit and researchers may miss the cues, particularly if they are from different cultures. We really need to create an environment where people feel they belong and know what they have to achieve to reach their goals.

Leaders and managers must be committed to change and seek out professional input. We had a specialist in group dynamics observe our university promotions committee. The committee functions much better after learning about group dynamics and about how to intervene and redirect conversations. In my work with the European Physical Society we made the rule that if there are no women nominees for a prize the prize will not be awarded. This obliges the research community to look for women candidates.

In physics we struggle with competitive and aggressive behaviour. Sometimes in the field it is considered a positive thing when you are aggressive about your findings and achievements. This is something that the scientific community should eradicate as it negatively impacts not just on women but also on many men who are not comfortable with this way of working. There has to be a consensus from the community and actions to move to a more inclusive culture.

Discussion: common themes

While the number of interviews is too small to be representative of the whole physics community, it is nonetheless interesting to explore common themes that emerged.

All of the interviewees were drawn into physics due to its intellectual challenges and due to enjoyment of exploring the forefront of scientific knowledge. Physics is a field of constant discovery and physicists find it rewarding to uncover novel phenomena and new knowledge of Nature. Exploration of the unknown is a strong pull factor towards physics. Interviewees raised the point that other domains may have their own powerful pull factors: the issue is not just that physics may push under-represented groups away. For example, some may be drawn towards fields in which they can have more immediate socio-economic impact.

A central theme in all interviews was the importance of community, networking, and mentoring in physics. Early career physics researchers benefit from networking and mentoring on how to progress from doctoral research through the mandatory postdoctoral phase towards a permanent post. Senior researchers view such guidance of early career researchers as a primary part of their roles and celebrate the success of their junior colleagues and collaborators. However, high competition for permanent academic posts penalizes those from under-represented groups who have less access to such advice and mentoring.

Interviewees noted that the culture of the field has progressed over the last few decades. For example, overt sexism of language and behaviours would no longer be tolerated. Yet more subtle behaviours and biases still persist, and under-represented groups may find themselves excluded from informal knowledge such as unwritten expectations for appointments or promotions.

Interviewees suggested a number of actions that the physics community could take to improve diversity and inclusion across all career stages. No interviewee advocated that any single action would on its own qualitatively change the situation: a range of actions was suggested to attract and retain more women in physics research. Many of the approaches proposed, such as mentoring of researchers from under-represented groups, have consistently been shown to be effective. However, physics researchers may not know what approaches to improving diversity and inclusion are evidence-based.

The interviews highlight a number of interesting questions for future research. Culture surveys are increasingly used to explore the working environment in physics, both by research funding councils and by gender action plans. Such surveys are valuable in identifying issues and areas where the culture could be improved, but most surveys span diverse groups. For example, university or national level surveys will typically span all sub-fields in physics and hence give limited information about the culture of specific sub-fields of physics. As this chapter demonstrates, interviews are a valuable and complementary tool to obtain detailed and nuanced understanding of the highly complex physics community.

Modern physics research is highly diverse in its nature, ranging from theorists working in very small groups to experimentalists in areas such as particle physics working in collaborative teams with thousands of members. Relatively little data is available on how the gender representation varies by sub-fields in physics but empirically it is noted that women are less represented in theoretical areas and more highly represented in experimental areas, particularly those that interface with biosciences and medicine. Further in-depth interviews and comparative research could give rich insights into the culture and environment differences between sub-fields of physics.

Interviews would also be important in exploring intersectionality of gender in physics with characteristics such as race and ethnicity, socio-economic background, sexual orientation, disability, caring responsibilities, and religion. An important part of intersectionality in physics relates to the global nature of the research field and the requirement for mobility. Researchers are expected to move between countries for postdoctoral research and may not be able to obtain permanent academic posts in their home countries. It would be interesting to explore how this culture affects retention of under-represented groups as well as the perception of inclusion among those working outside their home countries.

Appendix

Interview-structure guide

Following the principles of semi-structured interviews, we will use an interview-structure guide to ensure organization but flexibility of the interviews. The guide includes a set of proposed interview questions. Although these questions are prepared beforehand to guide the conversation, the interview will prioritize open-ended questions and encourage two-way communication to explore in depth. All the interviews will be recorded via Zoom and the recordings will later be used to produce the written accounts.

Type of interview: semi-structured in-depth interview.
Interview length: 45–60 minutes.
Interviewees: 4–5 interviewees, reflecting diverse career paths and career stages, and diverse protected characteristics (gender, ethnicity, nationality, and so on).
Interview channel: videoconferencing via Zoom.

Potential interviewees will be invited to participate by the interviewer. Once the date/time is fixed, an interview reminder will be sent over to the interviewee with the consent form 1–3 days before the interview. The transcript of the interview and the proposed final narrative for the book chapter will be passed to the interviewee to correct errors and give approval. Access to the full transcript and recording will be limited to the book editors only.

Interview questions

Career background

What is your scientific background? Why did you choose to become a scientist? Did a role model influence your decision to become a physicist?
In which area of physics do you work? Can you give a general overview of your research?
How did you choose your specific field of research?
Do you come from an academic family? How does your family regard your career choice?
Are there challenges in balancing your physics research career with your private life?
Why do you enjoy physics?

Career support

During your career, have you been mentored or supported? What is your own approach to mentoring young researchers?
What advice would you give to somebody starting a PhD in physics?

What is the best career advice you have been given?

Obstacles and challenges

What would you consider to be your biggest achievement? And your biggest failure?

What obstacles have you had to overcome in your research?

What kind of prejudices, if any, did you have to face? How did that make you feel?

Have you ever doubted your abilities as a scientist? How did you handle these situations/feelings?

Did you ever have the impression that it would be easier/harder if you were a different gender?

Gender equality

Do you feel that there are gender differences in your research community/ environment?

Have your views on gender equality in science changed during your career?

What would you consider to be the biggest challenges (opportunities) for women in physics?

What changes, if any, are needed for academic science to be more attractive to women and other under-represented groups? At what stages do you think that women are lost from physics?

Have you noticed positive changes over time in the perception and number of women physicists? Do you think that initiatives to get more women into physics are succeeding?

If physics had a better gender balance, how do you think this would affect research culture?

Do you think improved gender balance would have impacts on the science in your field (publications, but also organization of experimental teams and so on)?

Is there anything else you would like to tell us related to gender in physics?

11

A Room of One's Own: Photographs of Women Physicists in Their Working Spaces

Meytal Eran Jona and Sharon Diamant Pick

This chapter is dedicated to visualizing gender and physics. It includes photographs of women physicists in their chosen setting, be this their workspace, office, or laboratory. To this end, we published a call in several networks of physicists throughout Europe (for example, GENERA, a European network for advancing gender equality in physics; the International Union of Pure and Applied Physics; the European Physical Society; among others) in which we invited women who are physicists to share their own photos of their working space, where they felt the physics 'magic' happens. Following the photos, we also invited them to tell the photo 'story' in their own words.

Having women's voices heard is part of a feminist project that has been going on for decades whose aim is to correct an entire history of human existence in which women's voices were silenced, their experience was subordinated to that of men, and their history was hidden.

This chapter was conceived as a tribute to Virginia Woolf, one of the first feminist thinkers and mothers of gender theory. The name of the chapter, 'A Room of One's Own', is a tribute to her iconic book with this title, written almost a hundred years ago. The book made people aware of the resources required for women in order to be able to engage in writing. Borrowing from her, in this book, we think of the resources women need to become physicists. The 'room' of which Virginia Woolf spoke expanded and became, in the case of physicists, sophisticated top-of-the-line laboratories, state-of-the-art technological devices, as well as the human and financial resources that are required to do physics, in a space that is, today, more competitive, technological, and challenging.

The following photographs shared with us, as well as the story behind them, reflect an exciting mosaic of women physicists, each with their own story and unique life path. Apart from their common occupation, the women differ from each other in nationality, age, career stage, ethnic and religious origin, the field of research within physics, and more.

This chapter is a celebration of diversity. It exposes the many ways in which physics is studied, from fieldwork in Antarctica, to an office room, in many languages and countries, by doctoral students and senior professors in small teams and huge collaborations, in Europe and beyond.

We invite you to physicists' personal working spaces, to a collection of photographed moments, of diverse women, who do physics, each in her own unique way.

Sohyun Park

Current position: Visiting Researcher, Boston University, US
Country of birth: South Korea
Field of research: heavy ion physics

Figure 11.1: In Sohyun Park's office as a Postdoctoral Fellow at CERN

Photo taken by: Aleksas Mazeliauskas

This picture shows the most likely way you will find me if you pass by my office on most days. My door is usually open, and I am always looking forward to the next interesting chat. In theoretical physics, a lot of research projects are initiated by informal and unexpected discussions. My advice to future scientists is to start talking to your fellow physicists on any topics you are interested in.

Beena Kalisky

Current position: Professor, Bar-Ilan University, Israel
Country of birth: Israel
Field of research: experimental condensed matter physics

Figure 11.2: Beena Kalisky and her daughter, Roni, in her lab at Stanford University during her postdoc

Photo taken by: Lan Luan

This picture was taken at Stanford, during the winter break, when daycares were off. I was curious about the results of my experiment but also wanted to spend time with my family. That day my husband took care of our 4- and 2-year-old sons, and my daughter came to the lab with me to do some cool science. I think she had a great time, and so did I. My advice is to do whatever works for you, set your own rules, find your balance, and not give up on any of your dreams.

Betti Hartmann and Suparna Roychowdhury

Betti Hartmann

Current position: Associate Professor, Department of Mathematics, University College London, England
Country of birth: Germany
Field of research: applied mathematics

Suparna Roychowdhury

Current position: Assistant Professor, Department of Physics, St Xavier's College Kolkata, India
Country of birth: India
Field of research: astrophysics and cosmology

Figure 11.3: Betti Hartmann (left) and Suparna Roychowdhury (right) at the observatory of St Xavier's College in Kolkata

Photo taken by: Tanaya Bhattacharyya

This photo shows the interaction between two women scientists from very different backgrounds: Betti Hartmann, an applied mathematician at University College London, raised and educated in Germany, and Suparna Roychowdhury, an astrophysicist and teacher at St Xavier's College Kolkata, born and raised in India. The two, and their student teams, are collaborating on a joint research project in astrophysics. Women scientists from the Global South or India often have to overcome cultural restrictions and forms of discrimination that do not exist in Western societies when engaging on a scientific career. This photo is a reminder that a fight for equality of women should include all women.

Eija Tuominen

Current position: Adjunct Professor, Helsinki Institute of Physics, University of Helsinki, Finland
Country of birth: Finland
Field of research: particle physics and instrumentation

Figure 11.4: Eija Tuominen in her lab at the University of Helsinki

Photo taken by: Meytal Eran Jona

My office door is always open to colleagues and students! I enjoy sharing and discussing ideas, views, and opinions – or just meeting people for informal chats. In physics and science, this kind of communication, interaction, and networking may lead to new successful research projects and great results. I advise future scientists to trust yourselves, find companionships, and change the world!

Chiara Mariotti

Current position: Senior Researcher, INFN Torino, Italy, Adjunct Professor in Physics, Boston University, US
Country of birth: Italy
Field of research: experimental high-energy physics

Figure 11.5: Chiara Mariotti at CERN

Photo taken by: Bennett, Sophia Elizabeth, ©CERN

I like this picture very much, because it reminds me of the greatest memory of my professional life: the discovery of the Higgs boson. This picture shows me at the blackboard explaining how the Higgs boson can be produced and then decay at the LHC collider. Searching for the Higgs boson was a great passion that absorbed me for about 15 years and culminated in its discovery in 2012. I love to talk about it and teaching how to do research to young people is very rewarding.

Nadya Mason

Current position: Professor and Dean of the Pritzker School of Molecular Engineering at the University of Chicago
Country of birth: US
Field of research: experimental condensed matter/quantum materials

Figure 11.6: Nadya Mason in her research laboratory

Photo taken by: L. Brian Stauffer

This picture shows me in my low-temperature physics laboratory. I chose to study physics not only because I love the way it explains so many aspects of our day-to-day world, but also because I really enjoy playing around in the lab. It's like an adult sandbox – but where we get to learn cool and maybe useful things about the world at the end of the day.

Raquel Gómez Ambrosio

Current position: Assistant Professor, University of Turin and National Italian Institute for Nuclear Physics (INFN), Italy
Country of birth: Spain
Field of research: theoretical particle physics

Figure 11.7: Raquel Gómez Ambrosio in her office at the University of Milan Bicocca

Photo taken by: Setareh Fatemi

Here I was going back to a problem that has been in my mind for a few years now: trying to understand the Higgs potential. This is one of the current challenges in the field of high-energy physics. I like this picture because it portrays one of those moments when tackling a problem is actually a lot of fun. While there are times when one faces uncertainty and despair, there are also those 'Eureka' moments when the pieces seem to make sense together, make everything else worth it.

Hagar Landsman Peles

Current position: Senior Staff Scientist, Department of Particle Physics and Astrophysics, Weizmann Institute of Science, Israel
Country of birth: Israel
Field of research: astro-particle physics

Figure 11.8: Hagar Landsman Peles at 'ARA2', 3 km from the Amundsen–Scott South Pole Station, Antarctica

Photo taken by: Hagar Landsman/ARA

Here I was calibrating the 'ARA-2' detector deployed in the Antarctic ice at a depth of approximately 200 meters. It is very challenging to run an experiment at the South Pole.

In this photo I was using the handheld radio device to communicate with my colleague who sat in the control room near the station. During measurements we had to keep radio silence, and I will never forget the magic and serenity of being in absolute silence surrounded by pure whiteness and nothingness.

I was selected to be a part of the 'on-ice' team three times; each time I was away from home for a month.

Eva Hackmann

Current position: Senior Scientist, Center of Applied Space Technology and Microgravity (ZARM), University of Bremen, Germany
Country of birth: Germany
Field of research: gravitational physics (general relativity, astrophysics, relativistic geodesy)

Figure 11.9: Eva Hackmann in ZARM at Bremen University

Photo taken by: Jasmin Plättner

My scientific career is very closely tied to the drop tower in Bremen seen on the large canvas in the background. I deeply appreciate the prospering and stimulating environment with a wide range of scientists offering different views on space-related topics, discussed at the desk shown in the picture. For me it is a permanent scientific home, and without it I certainly would have left academia long ago.

Yasmine Amhis

Current position: Senior Researcher, Laboratoire de Physique des 2 Infinis Irène Joliot–Curie (IJCLab), France; Physics Coordinator of the LHCb experiment, CERN
Country of birth: Algeria
Field of research: flavour physics

Figure 11.10: Yasmine Amhis in the Physics Coordinator Office at CERN

Photo taken by: Cindy Denis

In this picture I am sitting in the Physics Coordinator Office. It is both a privilege and a responsibility to occupy this space, but one can see that I am happy to fulfil this role. In the background there are my dear notebooks that I have been keeping for many years – I record my thoughts, work, and notes from all the discussions that I have with my collaborators. One can also find in this office besides my physics books an art book and art supplies, which I turn to when I need a change of perspective.

Anastasiia Zolotarova

Current position: Permanent Researcher, CEA/Irfu/DPhP, Institute of Research into the Fundamental Laws of the Universe, Particle Physics Division, France
Country of birth: Ukraine
Field of research: particle physics

Figure 11.11: Anastasiia Zolotarova at Laboratoire de Physique des 2 Infinis Irène Joliot–Curie (IJCLab)

Photo taken by: Dominique Longerias – IJCLab/CNRS

I chose to show the experimental facility of the lab – the heart of our research group. It is a cryostat, capable of cooling the samples to temperatures of a few tens of millikelvins – close to absolute zero. This type of detector is very effective for the search for rare events, especially the search for neutrino-less double beta decay, which is the main topic of my research.

The work on this ultra-low temperature cryostat is demanding, but fascinating. Dealing with these complex technologies became my everyday life, which I enjoy and appreciate.

Cameron Norton

Current position: PhD Student, New York University, US
Country of birth: US
Field of research: theoretical cosmology and particle physics

Figure 11.12: Cameron Norton in her home office in New York City

Photo taken by: Sophie Goodman

I chose to show me at my home workspace because this is the room I truly feel is my *own*. I think that my personalization of this space is contrasting to the canonical picture of a physicist's office, which is male-centric. I fill my office with plants that bring life into the environment, and a quote that emulates the playfulness and magic that is the reason why I am drawn to physics.

Marie-Hélène Schune

Current position: Research Director, Laboratoire de Physique des 2 Infinis
Irène Joliot-Curie (IJCLab), Orsay, CNRS, France
Country of birth: France
Field of research: particle physics

Figure 11.13: Marie-Hélène Schune in her office at Laboratoire de Physique des 2 Infinis Irène Joliot-Curie (IJCLab)

Photo taken by: Luc Petizon – IJCLab/CNRS

It is a place of life: ideas are born there, they evolve there, often disappear ... but not all of them!

Reinhild Fatima Yvonne Peters

Current position: Professor of Particle Physics, University of Manchester, England
Country of birth: Germany
Field of research: experimental particle physics

Figure 11.14: Reinhild Fatima Yvonne Peters in her office at the University of Manchester

Photo taken by: Alessandra Forti

The picture shows myself in my office, with all my usual mess of papers and notes, my laptop with a lecture on my research topic of the top quark on the screen, and my 'assistant' Dr Genghis Hahn. Thus, this photograph includes the two passions of my life: top quarks and chickens.

When the pandemic was in full swing, we had to make videos for our lectures. I got myself that little rooster hand puppet to make the videos a bit more fun. My lectures are on particle physics, which, historically, developed out of nuclear physics. Otto Hahn was a nuclear physicist, a Nobel laureate (and German, like myself). Hahn means 'rooster' in German, and thus came the idea to have Otto Hahn's 'great grandnephew' in my lectures.

12

Conclusion

Pauline Leonard, Meytal Eran Jona, Marika Taylor, and Yosef Nir

The chapters of this book have built up a picture of gender and physics in academia as deeply and unequally structured through historical and organizational cultures. Organizational policies, which fail to effectively challenge the persistence of gendered inequalities, translate and sustain the imbalance, which is further operationalized through group and individual practices of discrimination, sexism, and microaggression. We have shown how these gendered structures, cultures, and practices are ubiquitously experienced by women physicists from diverse backgrounds, career stages, and ages. Building our picture of the disparity in the academic physics workforce has required drawing together critical analyses of different national and international contexts and policy initiatives, theoretical perspectives, and debates seeking to explain gender inequalities, and the personal life stories and images of physicists themselves. Weaving together these varied contributions from a distinguished international team of physicists and social scientists has enabled exploration of the complexities of European academic landscapes to fulfil four central aims: to demonstrate the loss to physics as a discipline through the marginalization of women; to address the question of why the under-representation of women in physics remains endemic and slow to change; to demonstrate that strategic leadership, evidence-based policies, and successful role models are essential for change to be effective; and to offer recommendations for policy to achieve a sustainable culture shift in academic physics. We now turn to consider these in turn.

A significant loss

Why is it important that the discipline of physics in academia is more gender-diverse? The contributions in this collection demonstrate how the answer to this question lies in three interconnected fields: physics, academia, and society. Within the discipline, the quality and achievements of physics research will benefit from a larger participation of women in two ways. First is a rationalist argument of capacity: having more women involved in physics research has the potential to double the pool of talent from which both academia and the private sector can draw physicists. Such a situation will mitigate the shortage of physicists expected in the future and, even more significantly, will diversify how excellence in physics is (to be) understood.

The second factor is that increased gender diversity will bring new ways of thought, new priorities, new questions, and perhaps new answers into physics. Here, by definition, it is practically impossible to give concrete examples of these novel developments, because it is precisely the absence of women's voices from current physics that narrows down the scope of possibilities for questions asked and answers given. Of relevance here is that a marked correlation is identified between the presence of women in top management teams and organizational performance (Desvaux et al, 2017; Jarboe, 2019). Women in leadership positions improve organizational performance in terms of vision, motivation, accountability, leadership, work environment, and values (Desvaux et al, 2017). Further, enhancement to organizational performance requires a diversity of leadership styles: not only gender, but also diversity brought through race and ethnicity, cultural and social background, sexuality, physical and mental ability, age, and so on. A fully diverse workforce is also essential for creating a more tolerant and inclusive environment (EHRC, 2019).

Finally, the loss of women to physics raises fundamental questions of social justice. We have seen how many of the factors that lead to the under-representation of women in physics arise from broader societal discourses about gender/science. The contributions in this collection have evidenced how the culture of physics is a masculine field, and that the perception of the ideal physicist is as a brilliant 'Hercules' male, for whom physics is the sole, all-encompassing passion in life. For men, the workplace identity is often one of a 'big ego' (Hasse and Trentemøller, 2008). At the same time, a strong and contradictory discourse prevails: that academic decisions and successes are based solely on meritocracy. This leads the academy to being blind to the various obstacles and difficulties faced by women (see Eran Jona and Nir in this book). Consequently, when pursuing an academic career, women face an unequal playing field of opportunities and experiences. Providing equal opportunities to women is thus a matter of human rights, and society as a whole will benefit from correcting the situation.

Change is slow

Despite the power of these arguments, the book reveals the slowness of change. Despite the introduction of multiple policy initiatives to improve the recruitment and career attainment levels of women in physics – and science, technology, engineering, and mathematics (STEM) subjects more generally – it is clear from our contributions from women physicists across a broad range of European contexts that the under-representation of women remains pervasive. Indeed, research shows that if current hiring practices and attrition rates are maintained, the fraction of women in physics will remain below 30 per cent for at least 60 years (Kewley, 2021). Even in the most optimistic alternative scenarios, gender parity is unlikely to be achieved for another 25 years. We have shown that 'pipeline stress' – or the attrition of women from the discipline – is a key cause, with women leaving the profession at rates nearing three times the rate as men at the postdoc level and almost double the rate at associate professor level (Kewley, 2021). However, the chapters in this book also argue that a sole focus on attrition as a *cause* obfuscates how this is the *result* of the complexity of cultural issues which can make workplaces hostile environments for women and other marginalized groups. In their chapters, Eran Jona and Nir in this collection provide important evidence on the complexity of factors impacting the postdoctoral phase, demonstrating the ways in which gender power operates in multiple and often hidden ways in the labour market, within physics as an academic field, in the family, and in the social norms and expectations of society, all creating barriers to women's ability to continue for a postdoc and, later, to pursue an academic career in physics.

A major contribution of the collection to improve our understanding of this situation is the evidence that shows that gender relations are not fixed entities but are continuously and repetitively made in the routine practices of everyday life (Butler, 1990; Tauber, 2020). Further, the biographies of women physicists underline how gender must always be understood in intersection with other social identities, such as inter alia nationality, race and ethnicity, social class, age and sexuality, and so on; all of which can combine to be sources of additional marginalization. Unequal power relations need continual maintenance, such that sexism becomes a normal, rather than an aberrant, aspect of the workplace: a 'business-as-usual' form of active discrimination that women confront every day (Breen and Meer, 2019; Leonard, 2021). Contributions by Brage and Drew, Lund and Aarseth, Eran Jona and Nir, and Sekula all reveal how it is the micro-practices of working life which feed the macro-level inequalities of a gendered society. The normalization of gender inequality can help to explain the slowness of change: an embedded feature of 'business-as-usual' discrimination is *social desensitization* (Breen and Meer, 2019) which increases the threshold of what counts as 'real' sexism, racism, and so on. Focus often concentrates on high-profile occurrences, such as leading scientists complaining that 'girls' in labs tend to fall in love with their male colleagues and cry when criticized (Nature, 2015) or denying that physics suffers from male gender bias and criticizing affirmative-action policies (Castelvecchi, 2018). Within academic institutions, formal action against breaches of equality, diversity, and inclusion (EDI) is rare, as complaints tend only to be made for particularly overt and witnessed instances of sexism and discrimination – and not always then. This can lead to a 'smokescreening' of the daily microaggressions by which sexism and discrimination are thoroughly embedded in, and operationalized through, the infrastructures of academic departments.

The book also demonstrates how, in recent years, we have seen the introduction of a raft of policy initiatives in academic physics departments across Europe. These aim to better understand the factors contributing to women's under-representation and tackle the systemic issues which work to exclude women from the discipline. However, what we learn from accounts of these policy programmes is that, while the focus on institutional policies such as recruitment, promotion, parental leave, and so on is important, we must also tackle the complex, subtle, and often obscure ways in which bias and discrimination operate. To change cultures of marginalization and oppression, leadership which fully acknowledges the overt and covert ways in which these are administered is essential.

Achieving change

The various chapters in this book underline the decisive role that the *leadership* of the institution has in formulating and implementing policies to promote gender equality and diversity. Leadership is a major factor in organizational transformation and is critical to successful equity and diversity efforts (Colwell, Bear, and Helman, 2020). As demonstrated by Thomas Berghöfer, Helene Schiffbänker, and Lisa Kamlade in Chapter 7, as well as through the discussions held in our Conference on Gender and Physics held in 2019, change happens when an institute has a solid leadership that sets the agenda and ensures the implementation of initiatives and policy to enhance diversity and inclusion. Across academic and business contexts, we see that that organizations are more successful in promoting equality when leaders are personally engaged in problem solving, when close relationships are established with women and minorities, and when the leadership want to be perceived as fairer (Dobbin and Kalev, 2016). In organizations where the management did not support the equality agenda seriously and fully, change did not happen, again clearly shown in Chapter 7.

Organizational culture is thus also critical to achieving gender equality. The importance of changing institutional culture as a tool for organizational transformation in STEM fields has been powerfully stressed by scholars in the field (Bohnet, 2016; Colwell, Bear, and Helman, 2020) and the chapters in this collection build on these findings to argue that, within physics departments or research institutions, we should identify and remove old habits and non-effective organizational procedures that introduce barriers to gender equality. These include, for example, hiring procedures and criteria that impose extra hurdles on women candidates, unequal distribution of leadership roles within departments and organizations, and the unequal distribution of resources, all of which fuel a culture of male privilege and entitlement. In contrast, it is essential to work proactively to enhance women's visibility: if being a 'woman physicist' is to be normalized to 'being a physicist' then women must be highly visible in leadership positions; be invited in significant numbers to give plenaries at conferences, colloquia, and seminars; and be celebrated by the discipline and their institutions for their achievements.

To succeed in reinforcing cultural change in academic physics, challenging the dominance of masculinity, improving the gender balance at all career stages, and constructing departments which are open to people of diverse backgrounds, an organization must develop *policies* to retain the quality of their employees and manage the talent pool. In Chapter 1, we outlined a range of feminist perspectives and argued that these differing assumptions underpin the design of different policies aiming to tackle issues of inequality. Yet while the chapters in this book have drawn variously on structural and liberal approaches, and occasionally elements of radical, materialist, and postmodernist feminism, to analyse the experiences of women physicists across Europe, the discussion of the policies which have been introduced reveals that these tend to be dominated by liberal feminist solutions. We have seen that policies tend to focus on improvements such as the provision of maternity/parental leave, childcare; and 'family-friendliness'; collecting data on gender bias and inequality to improve awareness and understanding; and strategies to cope with discrimination, sexual harassment, and bullying. All of these 'soft' solutions aim to remove barriers to success and enable women to compete as individuals on a more 'level playing field'. Yet while these policies are felt by our contributors to have benefitted women's careers and improved the cultures within academic physics departments, the slowness of change and the consistency of under-representation begs the question of whether they are effective in dismantling the deeply embedded structures of inequality which exist. More fundamental solutions – such as targets and/or quotas in both recruitment and leadership, women-only hiring campaigns, and dismissal of sexual harassers and bullies – may be needed to radically transform the structures of power. However, following a postmodern approach, the chapters in this book also show how a major step towards effective gender equality in higher education is that each organization embraces diversity and inclusion policies that fit their specific needs, based on national context, societal culture and norms, and individual institutional and disciplinary features. Different institutions have different goals and missions, values, cultures, and resources, and this institutional context can impact the efficacy of policy. Studies have shown there is no one-size-fits-all solution, policy, or practice that will perfectly fit the needs of all institutions (Colwell, Bear, and Helman, 2020); therefore, each institution should design their own gender-equality plan, drawing on the evidence of *best practice* and with clearly identified and agreed outcomes.

In this book, we have demonstrated how physicists are working together to develop new programmes to bring about change. We reviewed three projects which over the last ten years have gained traction within the physics community itself: the GENERA project, which includes nearly 40 European countries and beyond (Chapter 7); the COST project within the string community (Chapter 8); and the JUNO project in the UK (Chapter 9). All of these initiatives were 'bottom-up': growing out of the physics community itself, and later receiving support from the universities, professional physics associations such as the European Physics Society, the International Union of Pure and Applied Physics, and the Institute of Physics, as well as the European Commission. All the programmes shared the common goal of promoting gender equality in academic physics. The lessons learned and best practices were debated at a workshop held at the Weizmann Institute of Science in October 2019, entitled 'Promoting Gender Equality in Physics: Barriers and Opportunities'. It was this workshop which led to the birth of this book. Participants included leading physicists from all around Europe, social scientists who specialize in gender studies and organizational change (including a few leaders from the US), and practitioners who lead initiatives to enhance gender equality.

Drawing from the conclusions and recommendations from these programmes, as well as the contributions in this book, we conclude by identifying the following principles for policy:

- Use social science methodologies to collect data and monitor trends in the recruitment, retention, and advancement of women to better adapt targeted interventions and to monitor their efficacy.
- Based on the data, design and embrace a strategic plan to identify targets for hiring, retention, and promotion to achieve gender equality.
- Adopt an anti-discrimination policy (code of conduct), for development of a sustainable and inclusive culture and the prevention and enforcement of the proper behaviours.
- Involve the whole community (men and women) in finding solution(s) inclusive for all genders, recognize diversity and intersectionality, as well as new family forms.
- Celebrate and enhance the visibility of women physicists, especially women from minority and under-represented groups, to increase the number of role models.

- Implement additional measures to support women in the postdoctoral phase, such as mentoring and financial support for those with caring responsibilities.

Closing thoughts

Recent global events have raised the profile of gender equality across multiple national and social contexts, and active movements for change are becoming stronger and more numerous. It is becoming more and more difficult to ignore inequalities of power that facilitate and shape male domination in socio-economies; or defend long-standing hierarchies based on gender, as well as race and class. While women have made huge progress in education and the workplace during the past 50 years, entering fields which were previously closed to them such as medicine, politics, law, and business, the gains in STEM education and careers remain depressingly slow. While change clearly needs to be made in early years education when the foundation for a science career is laid, it is in the academy that scientists and physicists are made. If women physicists see that there are few people 'like them', perceive that the culture of academic physics is hostile, and that different (higher) standards are applied to them, then why would they stay? It is high time for the academy to take a hard look at their biases, behaviours, policies, and practices and to lean into creating a fairer and inclusive environment for physics.

References

Bohnet, I. (2016) *What Works? Gender Equality by Design*, Cambridge, MA: Harvard University Press.

Breen, D. and Meer, N. (2019) 'Securing whiteness? Critical race theory and the securitization of Muslims in Education', *Identities*, 26(5): 595–613.

Butler, J. (1990) 'Performative Acts and Gender Constitution', in S. Case (ed) *Performing Feminisms: Feminist Critical Theory and Theatre*, New York: John Hopkins University Press, pp 270–82.

Castelvecchi, D. (2018) 'CERN suspends physicist over remarks on gender bias Nature', 1 October, Available from: https://www.nature.com/artic les/d41586-018-06913-0

Colwell, R., Bear, A., and Helman, A. (2020) *Promising Practices for Addressing the Underrepresentation of Women in Science, Engineering, and Medicine: Opening Doors*, Washington, DC: The National Academies Press. DOI: 10.17226/ 25585

Desvaux, G., Devillard, S., Labaye, E., Sancier-Sultan, S., Kossoff C., and de Zelicourt, A. (2017) *Women Matter*, McKinsey & Company, Available from: www.mckinsey.com/featured-insights/gender-equality/women-mat ter-ten-years-of-insights-on-gender-diversity

Dobbin F. and Kalev, A. (2016) 'Why diversity programs fail and what works better', *Harvard Business Review*, July–August.

EHRC (2019) 'Tackling racial harassment: universities challenged', Available from: www.equalityhumanrights.com/sites/default/files/tackling-racial-harassment-universities-challenged.pdf

Hasse, C. and Trentemøller, S. (2008) *Break the Pattern!*, UPGEM project, Estonia: University of Tartu Press.

Jarboe, N. (2019) 'Women's Leadership in Higher Education', Available from: www.advance-he.ac.uk/news-and-views/womens-leadership-in-higher-education

Kewley. L. (2021) 'Closing the gender gap in the Australian astronomy workforce', *Nature Astronomy*, 5: 615–20.

Leonard, P. (2021) 'Women getting in and getting on', *Nature Astronomy*, 5: 533–4.

Nature (2015) 'Sexism has no place in science', 522: 255, Available from: www.nature.com/news/sexism-has-no-place-in-science-1.17761

Täuber, S. (2020) 'Undoing gender in academia: personal reflections on equal opportunity schemes', *Journal of Management Studies*, 57(8): 1718–24.

Index

References to figures appear in *italic* type; those in **bold** type refer to tables. References to endnotes show the page number and the note number (231n3).